salamanders

OF THE EASTERN UNITED STATES

Larry Wilson, Whit Gibbons, and Joe Mitchell

salamanders

OF THE EASTERN UNITED STATES

The University of Georgia Press   Athens

A Wormsloe
FOUNDATION
nature book

© 2024 by the University of Georgia Press
Athens, Georgia 30602
www.ugapress.org
Designed by Mindy Basinger Hill
Set in Quadraat by Melissa Buchanan
Printed and bound by Friesens
The paper in this book meets the guidelines for
permanence and durability of the Committee on
Production Guidelines for Book Longevity of the
Council on Library Resources.

Printed in Canada
28 27 26 25 24 P 5 4 3 2 1

Library of Congress Control Number: 2024934536
ISBN 9780820365732 (flexiback)

In Memoriam

JOSEPH C. MITCHELL (1948–2019)

With sadness, gratitude, and lasting respect we thank Joe Mitchell for his outstanding contributions toward the writing and publication of *Salamanders of the Southeast*, on which this book is based. Joe's scientific research and work in the areas of conservation and education had a profound influence on the field of herpetology nationwide. His insights and guidance on salamanders and other amphibians and reptiles will be missed.

CONTENTS

People and Salamanders

all about salamanders

WHY SALAMANDERS?

Why should someone get excited about small, seldom-seen animals? For one reason, being "out of sight" is no justification for this spectacular group of animals to be "out of mind," especially in the eastern United States. The remarkably clandestine salamanders are a classic example of hidden biodiversity; they far outnumber all other eastern terrestrial vertebrate groups in species and often in population sizes. Many people do not know that eastern salamanders rival the region's other animals in their remarkable body coloration and patterns, including the dramatic displays of birds, snakes, and butterflies.

More than 130 full species and one hybrid complex are represented in this book, with more destined to be discovered through genetic research. In fact, the salamanders found in the eastern United States constitute 16 percent of the world total of 767 known species recognized in 2022. And they offer many opportunities both to appreciate nature and to understand the importance of even rarely seen animals to the ecology of local communities of plants and animals.

More salamander species live in the tropics than in temperate North America, but more major groups (families and genera) are found in the eastern United States than anywhere else. Most people never see salamanders because they are

above All salamander larvae are aquatic, have four legs, and develop large gills for oxygen uptake.

left The dwarf salamander is one of the few southeastern salamanders that has four toes on the hind feet instead of the usual five.

When you do find that fiery red, traffic light yellow, bright orange, or lichen green salamander, think about the fact that this group of animals has long outlived the dinosaurs.

so secretive, yet many are major players in aquatic and terrestrial food webs. Some salamander species reach such high population densities that without them their predators might suffer severe impacts, and their prey might rapidly spiral out of control. Unfortunately, the future of an unacceptably high number of salamander species is being placed in jeopardy as their habitat is destroyed or degraded by chemicals and land use alterations.

Even the scientific community has only recently begun to understand how these unique animals communicate with each other, what their relationships are with their environment and with other species, how they can live for so long on only a few meals a year, and why many cannot survive without the full complement of aquatic and terrestrial habitats in the landscape. Indeed, we are only now learning how many species we really have in the eastern United States and how they are related to each other genetically. We are learning how they respond to loss of vegetation cover, newly emerging diseases, and input of agricultural chemicals, usually to their detriment. We are also learning how to manage their populations in urban and suburban areas to help ensure that they persist in concert with human activities.

Salamanders may be useful environmental sentinels if only we pay attention to what they are telling

The lichen-like color and pattern on green salamanders provide camouflage when their background is covered with lichen.

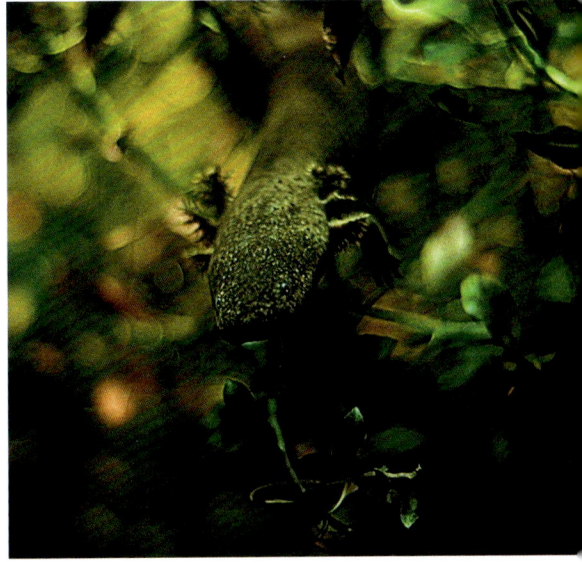

This greater siren illustrates the permanently aquatic, elongated salamander body form with only two front legs and gills.

us. The first salamanders in the fossil record date back about 230 million years, 40 million years before the appearance of dinosaurs. The salamanders outlived this spectacular group of reptiles by surviving the events that caused the dinosaur extinctions some 65 million years ago. Unfortunately, humans are doing things to the planet that are causing the extinction of salamanders far more rapidly than they can be replaced by evolutionary processes. Amphibians as a whole are declining at an alarming pace. If we do not change our environmental attitudes, humans may very well become the agents of the next great extinction of much of the animal life on Earth—the sixth such event in the planet's history (the fifth being the dinosaur extinctions).

Are amphibians today serving as indicators of the sixth great extinction? Most of the salamander extinctions are occurring in the tropics, but one salamander—the Bay Springs salamander (*Plethodon ainsworthi*), a woodland salamander species known from only two individuals caught in Mississippi in 1964—is thought to have become extinct within the last 60 years. Although others may be in jeopardy as a result of small, localized geographic ranges and unregulated environmental impacts, most eastern salamanders seem to be hanging on, and some are still doing well. What our generation does in the next decade or two

Salamanders, such as this terrestrial Yonahlossee salamander from the southern Appalachian Mountains, are classic examples of hidden biodiversity.

will determine how many of our existing salamander species will be around for future generations to study and appreciate.

This book is intended to inspire an appreciation of salamanders through the information it provides and the magnificent photographs taken by numerous herpetologists. Salamanders will not show themselves to you. You must find them by taking a hike or tour to see them in their native habitats or attending a nature class that has a selection of species on hand in captivity. Go out on nature walks in the daytime and turn over logs and rocks (replacing them afterward, of course) to see what lies underneath. Go out with a naturalist who knows how to find salamanders at night or during a rain, when they are most active. We encourage you to learn how to appreciate these amphibians and to value them in the same way that birds and other more visible animals are valued. And when you do find that fiery red, traffic light yellow, bright orange, or lichen green salamander, think about the fact that these animals crawled the earth before the dinosaurs, and that we humans may be edging many of them close to extinction. We hope that you will support efforts to protect and manage habitats everywhere so that our grandchildren and their grandchildren will have opportunities to appreciate the aesthetic beauty of these charismatic, seldom-seen amphibians.

DEFINING THE EASTERN UNITED STATES

Regions of the country can be identified from a variety of perspectives, including culture, political boundaries, geological or physiographic regions, and climatic or other environmental features. Clearly, human cultures or politics have little influence on the distribution patterns of salamanders, while the geological history and climate of a region have a great influence. But to continue the practical approach of using states as basic units for describing geographic ranges of species, we have chosen to define the eastern United States for the purposes of this book as those of the Southeast—Alabama, Florida, Georgia, Kentucky, Louisiana, Mississippi, North Carolina, South Carolina, Tennessee, and Virginia—with the addition of Connecticut, Delaware, Illinois, Indiana, Maine, Maryland, Massachusetts, Michigan, eastern Minnesota, New Hampshire, New Jersey, New York, Ohio, Pennsylvania, Rhode Island, Vermont, West Virginia, and Wisconsin. The defined region has an identifiable assemblage of salamanders distinct from those of the western states, whose biological character is shaped by different environmental factors and topographic features.

Amphibians are declining at an alarming pace. Ornate chorus frogs, for example, no longer occur in some parts of their range in North America.

BIOLOGY OF SALAMANDERS

Salamanders are members of the order Caudata ("caudate" means "tail") in the class Amphibia. Other amphibians include frogs and toads (order Anura), which lack tails as adults; and caecilians (order Gymnophiona), a group of legless, wormlike animals found in the tropics. By early 2022 scientists had described more than 8,720 species of amphibians worldwide, of which 816 are salamanders; 326 amphibian species live in the United States (116 frogs and 210 salamanders). The eastern United States, as defined in this book, is home to 120 species of salamanders. The first fossil salamanders known from North America were found in Arizona and date back to the early Jurassic period, some 200 million years ago. Thus, the modern salamanders in the eastern United States are descendants of an ancient group of vertebrates.

The word "amphibian" means "both lives," indicating that most amphibians have both a larval stage that is aquatic and an adult stage that is usually terrestrial. With the exception of some of the salamanders and a few frogs and caecilians, all amphibians remain tied to water just as their ancestors were. Most salamanders either live in water permanently or return to it to lay eggs. The eggs hatch into aquatic larvae that grow for a time, usually weeks or months, before losing

DID YOU KNOW?

All salamanders are carnivorous. Most eat insects and other invertebrates, but some, such as dusky, spring, and tiger salamanders, will eat other salamanders, including smaller individuals of their own species.

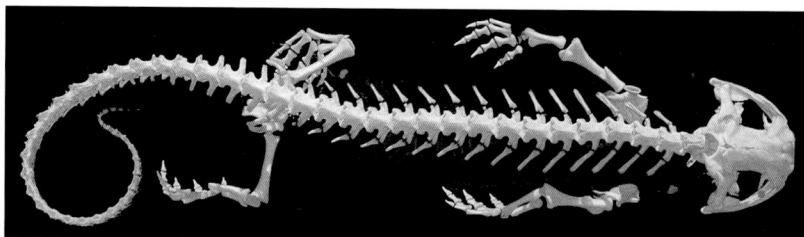

Salamanders have an elongated backbone, a long tail, and usually four limbs roughly similar in size.

their larval characteristics through metamorphosis. Salamanders have thin, permeable skin and must have special behaviors or adaptations to tolerate dry environments.

Salamanders differ from frogs by having an elongated backbone with 10–60 vertebrae, a long tail, and usually four limbs that are roughly similar in size. Frogs have only 5–9 vertebrae, a unique bone called a urostyle in their pelvis, and much enlarged rear legs used for jumping. All salamanders have a tail—usually a long one—a character that distinguishes them from caecilians with their stubby tail and frogs with no tail at all as adults. Salamanders also have a slender, elongated body and a distinct head.

Diversity of Salamanders in the Eastern United States
The 138 species of salamanders that inhabit the eastern United States belong to 17 genera within 7 families. Ten species are members of the family Ambystomatidae, commonly called mole salamanders because they spend much of their lives underground. The mole salamanders have a stocky body, a robust head, and four functional limbs.

The family Amphiumidae includes 3 species of fully aquatic salamanders called amphiumas, one of which reaches the longest total length of any salamander in North America, a whopping 41 inches. All have an elongated body, two pairs of tiny legs, lungs, and internal gills that are not visible on the outside. The only obvious difference among the species of amphiumas is the number of toes on each foot.

The family Cryptobranchidae has only one U.S. representative in the eastern United States: the hellbender. This large, prehistoric-looking salamander has a wide, flattened head; a robust body with wavy folds of skin along each side; and a paddle-like tail. Other members of this group live in the Ozark Mountains of Arkansas and Missouri, Japan, and China.

The most diverse group, not only of salamanders but of all U.S. vertebrates,

left Two species of newts in the eastern United States secrete poison from skin glands that deters would-be predators from eating them. This is a red-spotted newt.

right Thirty-seven species of salamanders in the genus *Desmognathus* live in rocky streams and adjacent moist forest habitat in the eastern United States.

above Mudpuppies like this one and its relatives the waterdogs live in permanent streams in the eastern United States. All are fully aquatic as adults and retain external gills to breathe in their aquatic habitats.

left Forty-two species of terrestrial salamanders in the genus *Plethodon* are found in the moist forests of the eastern United States. All lack lungs and do not have an aquatic larval stage.

right The ambystomatids, commonly called mole salamanders because they spend much of their lives underground, have stocky bodies and robust heads. Left to right: mole, marbled, tiger, spotted, and Mabee's salamanders.

is the family Plethodontidae, with 10 genera and 107 species in the eastern United States. A major anatomical feature unites all of them: they have no lungs. All their respiration occurs by gills or across the skin. The plethodontids have diverse body shapes, but all have four limbs, a long tail, a groove that runs from the nostril to the lip (nasolabial groove) on each side of the head, and movable eyelids.

The Proteidae include 7 species of mudpuppies and waterdogs, all of which are aquatic throughout life. They are paedomorphic (i.e., the adults retain larval traits) and have external gills, short limbs, and a tail that is laterally flattened.

The two members of the Salamandridae in the eastern United States are called newts. Unlike all the others, these salamanders lack costal grooves, and their terrestrial life stages have granular skin embedded with poison glands that make them distasteful to predators. Their brightly colored body or belly serves as a warning that they are toxic.

The Sirenidae includes 2 genera and 7 species. All of the sirens are fully aquatic, eel-like in shape, have external gills, and completely lack the rear pair of legs and pelvic girdle.

Most eastern salamanders are too small to bite humans, but some, such as amphiumas and hellbenders, can bite hard and even break the skin. Some, such as the newts, produce toxic secretions in skin glands that would make someone who swallowed or licked one sick. So be careful when picking up a large salamander, and don't swallow any of them.

Activity and Energy Salamanders, like all amphibians, are ectothermic, or "cold-blooded." Their body temperature approximates that of the surrounding environment because they cannot generate their own body heat from internal chemical reactions as mammals and birds do. Many salamanders are nevertheless active at surprisingly low temperatures. Tiger salamanders, for example, are well known for walking across snow to enter breeding ponds in winter and move about readily under ice cover. Their enzymes function best at low temperatures, and they are inactive in the warm summer months. The ectothermic lifestyle allows amphibians to spend less energy on keeping warm and more on growth and reproduction. Most require only a few meals per year to gain enough

left Some salamanders, like these red-cheeked salamanders, climb trees and vegetation to obtain prey on warm, moist nights. They are active on the ground surface only when temperatures are moderate and sufficient moisture is present.

right Salamanders store a considerable amount of energy in their tail in the form of fat. If the tail is broken, the salamander can regenerate another.

energy to survive, grow, and reproduce; and they can survive long periods without food. Salamanders do not hibernate in the usual sense. Some survive cold winter temperatures by seeking out refugia in forest soils or crevices below the frost line and becoming inactive.

Ectotherms are not as active and visible as most birds and some mammals; some species are above ground only a few times during an entire year, and only during night rains. About 90 percent of any woodland salamander (genus *Plethodon*) population remains underground at any given time, while 10 percent is above the ground foraging. Many species have highly seasonal activity patterns that are largely regulated by temperature and rainfall. Finding them thus requires knowledge of when and under what conditions they are likely to be active.

The majority of the eastern salamanders belong to the family Plethodontidae, all of whose members lack lungs. The adults of most species respire through their moist skin. They inhabit highly oxygenated streams or seepage areas or, if terrestrial, live in moist forests and remain inactive under surface objects or underground unless the ground is moist and the air humid from rains. The aquatic larvae of plethodontid salamanders and cave-dwelling adults use gills for respiration. All other salamanders have lungs as adults and gills as larvae, but also respire through the skin to at least some degree, and thus must live in humid environments.

Spring salamanders, found in eastern mountain streams and adjacent terrestrial habitats, are large enough to eat other salamanders, even large ones such as this red-legged salamander.

Salamanders as Predators All larval and adult salamanders are carnivores; that is, they eat other animals. Most adults consume a variety of invertebrates such as insects and their larvae, worms, millipedes, centipedes, and snails. Aquatic larvae eat zooplankton, mosquito larvae, and other invertebrates. Some salamanders will kill and eat larvae and adults of other species, and some even cannibalize their own kind. Large tiger salamanders can eat small mice. Amphiumas eat crayfish, tadpoles, and insects. Hellbenders prey heavily on crayfish and occasionally on small fish.

Terrestrial red-backed salamanders in Appalachian forests occur in such large numbers that their biomass may exceed that of all the birds and mammals that also occupy their habitat. They are responsible for an enormous amount of the energy that is transferred along the food chain from insects and other species that eat plants to the top carnivores in the ecosystem. Other salamanders can reach high densities in aquatic and terrestrial habitats as well, and all play a crucial role in energy dynamics in aquatic and terrestrial ecosystems. Salamander larvae link the two ecosystems because they transfer the energy they accumulated in the aquatic environment to the terrestrial environment when they metamorphose and emerge from wetlands.

Salamanders capture prey in several different ways. They detect potential prey visually or by smell. The size of the prey they can capture and swallow is in part determined by how wide they can open their mouth (gape), which in turn is related to their body length. Salamanders in terrestrial environments, such as mole salamanders, capture their prey by lunging a short distance with the mouth open; on contact they quickly extend the tongue forward and downward so that the back of the tongue contacts the prey and at the same time releases a sticky secretion that adheres to the prey. The tongue—and the insect or worm stuck to it—is quickly withdrawn into the mouth, which closes over it. Large prey are held in place by the teeth that line the upper and lower jaws. The plethodontid salamanders use a similar mechanism for feeding. Although their tongue is not connected in the front as it is in mole salamanders, they can protrude the tongue a short distance. The combination of a short lunge, open mouth, sticky protrusible tongue, and snapping the jaws shut is an effective mechanism for capturing prey.

Feeding in water requires a different strategy. Most salamander larvae and fully aquatic adults essentially vacuum their prey into their mouth. As the salamander makes a brief lunge at the prey, the floor of the mouth drops and the nostrils close, creating a vacuum that literally sucks the prey into the mouth. The mouth closes only partially to allow excess water to escape as it closes without losing the prey. Hellbenders are unique because they can drop one side of their jaw while keeping the other side closed. This allows them to capture prey on either side of the mouth rather than in front of it as most other salamanders do.

All salamanders are carnivores and obtain energy from a wide variety of invertebrate and small vertebrate animals.

Defense Mechanisms A host of other animals eat salamanders; some predators eat only one or two species exclusively. Large adult mud snakes, for example, eat only sirens and amphiumas. The large population size and biomass of many salamander populations offers predators an ample supply of food. In response, salamanders have developed a variety of ways to avoid becoming a meal.

Most commonly, both aquatic and terrestrial salamanders simply flee. Watch a slimy salamander quickly disappear down a hole as soon as you turn over the log under which it was hiding, for example; or try to catch a tiger salamander larva under the ice with a dip net and watch it avoid your efforts with twists and turns as it swims quickly away. Once caught, however, salamanders have other tricks to make the predator let them go. Large salamanders such as amphiumas and hellbenders can bite hard and rake their teeth across the predator's skin. Small salamanders can bite, too, but it is effective only with small predators. Many terrestrial salamanders that live under logs and rocks freeze in place when uncovered, taking advantage of visual predators' tactic of searching for a moving prey item; they tend to overlook a camouflaged, unmoving salamander. Small salamanders also rely on being distasteful or toxic. Newts secrete a powerful

Amphiumas and sirens are the primary prey of mud snakes in the Southeast. This mud snake is swallowing an amphiuma.

toxin from glands in their skin. This is why they can live in water with fish where species lacking toxins cannot survive.

Unlike the shelled eggs of reptiles, salamander eggs are protected only by several layers of viscous jelly, which makes them easy targets for predators such as predaceous invertebrates. Eastern newts are well known to eat eggs of tiger and spotted salamanders. Fish, giant water bugs, water tigers (an aquatic beetle larva), dragonfly nymphs, predaceous diving beetles, snakes, and wading birds are among the predators that eat salamander larvae, especially when the water is low and the larvae are concentrated. Marbled salamander larvae, which hatch early and are large by the time tiger and spotted salamander larvae hatch in the same pond, eat their smaller relatives. The tiger salamander larvae grow faster, however, and a couple of months later the tables turn, and the tiger salamander larvae are able to eat the marbled salamander larvae.

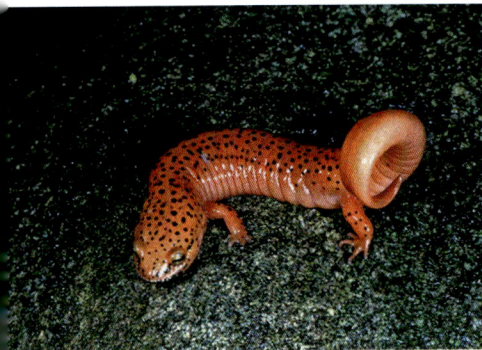

Some salamanders, such as the terrestrial juvenile eft stage of the red-spotted newt, are bright orange or red, warning would-be predators that they are toxic. A few species found in the same area, such as the red salamander,

Salamanders often distract a potential predator away from the head and body by waving or wiggling the tail. This red salamander is coiling its tail and exposing its undersurface.

mimic the toxic efts and are also brilliantly colored, and wary predators pass them by. Many of the woodland salamanders in the genus Plethodon secrete sticky, slimy mucus from skin glands that has been known to glue a predator's mouth shut. Although the mucus is nontoxic to humans, it is almost impossible to wash off and will remain on the skin for a day or two. Many salamanders escape predation by waving their tail to distract the predator's attention to that part of the body. Once the predator grabs the tail, the salamander can "autotomize" it, and the predator is left holding a wiggling tail while the rest of the salamander escapes. A new tail will grow to replace the one lost.

Behavior Unlike frogs, which vocalize to attract mates and defend territories, most salamanders lack the ability to make sounds. Sirens and amphiumas on rare occasions produce a low whistle or make clicking sounds; mole salamanders can whistle, click, or squeak; and some plethodontids can make soft squeaks. Little is known about these seldom-heard sounds, including why they are made, but they may be associated with defense or perhaps are a reaction to being captured and restrained.

Salamanders communicate primarily with chemicals and touch. Visual communication is limited because salamanders are generally underground, active at night, or in water with low visibility. All species produce pheromones—small-molecule chemical messengers designed to produce a specific response. Male woodland salamanders, for example, mark the boundaries of their small

This mud salamander is coiling its tail over its head to protect itself from a potential predator.

territories with pheromones from glands located posterior to the cloaca. The owner actively defends his territory, which is often limited to a rock or a portion of a small log, against intruders of the same or different species. Intruders know from the smell of the pheromone that the rock belongs to another salamander. The nasolabial groove on the front of the snout picks up the pheromone and transfers it to the nasal cavity. Sensory neurons transmit the signal from there to the brain, where it is interpreted. Intruders either avoid the territorial male or engage in a battle in which biting and wrestling determine the winner; usually it is the owner. Chemical and tactile communication also occurs between males and females as described below. Salamanders' chemical communication system is so sensitive that individuals can recognize other individuals, including their own brothers and sisters. Researchers are currently investigating the complex social systems of salamanders and the possibility that they may have short-term memory and cognitive thinking.

left Salamanders fight for several reasons, including territorial defense, capturing prey, and courting females. These two Black Mountain salamanders are locked in combat.

below Salamanders communicate largely by chemicals and touch during courtship and when evaluating potential intruders. The nasolabial groove on the front of the snout picks up pheromone molecules and transfers them to the nasal cavity, where they stimulate sensory neurons that send messages to the brain for interpretation.

Reproduction Salamanders accomplish reproduction with an amazing variety of behaviors that range from simple external fertilization of eggs to complex and elaborate rituals in which the male entices the female to pick up a packet of his sperm (i.e., a spermatophore) with her cloaca. The eggs are then fertilized internally. Some species lay eggs in water and have an aquatic larval phase; others are terrestrial and completely skip the aquatic phase.

Courtship in the mole salamander group (Ambystomatidae) is simple and fast-paced. Males arrive at wetlands before females and deposit rows of spermatophores (a mucus package of sperm). Females arrive days or even weeks later. A male approaches a female and nudges her body and cloaca. He directs her to straddle a spermatophore by walking backward in front of her, and with additional nudging moves her to position her cloaca over the spermatophore so that she can pick up the sperm packet. Fertilization is inter-

Egg follicles enlarge prior to fertilization and egg laying. The presence of yolking eggs in the belly of a female is an indication that the egg-laying season is about to begin.

nal, and the female then lays her eggs along grass blades or stems of other vegetation. Adults do not provide parental care. They leave the pond, and the eggs mature and hatch into aquatic larvae several days to weeks later. Marbled salamanders use a variation on this theme that allows them to produce young in the absence of water. Males deposit spermatophores on land during fall rains and entice females to pick up the sperm packets. Females deposit their eggs in loose clusters under logs and debris in the dry pond basin and stay with them until winter rains fill the pond with water. The eggs hatch shortly afterward, and the females make their way back to underground retreats in the forest.

Our knowledge of courtship in amphiumas comes from one observation in an outdoor enclosure of a female that glided over a male lying still in shallow water and positioned her cloaca opposite his. A sperm packet was apparently transferred during the contact. Following internal fertilization, female amphiumas lay eggs in pockets in the substrate, and these hatch into aquatic larvae. Mudpuppy courtship behavior is unknown. The male produces a spermatophore that is somehow transferred to the female, and fertilization is internal. Females provide parental care for their eggs.

Although external fertilization is the usual mode of reproduction among frogs, it is unusual among salamanders. Biologists postulate that fertilization in sirens is external but do not know the specifics because it has never been described. Hellbenders also reproduce by external fertilization. A male entices a female to enter a chamber under his rock in the stream, and she lays long strands of eggs on the underside of the rock. He then moves over them and extrudes white, ropy masses of sperm that fertilize the eggs. An additional oddity with hellbenders is that the male, not the female, remains with the eggs and protects them from potential predators. The eggs hatch into larvae with gills.

left Reproductive behavior in plethodontid salamanders such as these red-legged salamanders is tactile, chemical, and visual. The male rubs his chin on the female to transfer pheromones that will make her receptive to mating.

below The female Yonahlossee salamander arches her tail when the male straddles her and rubs his mental gland on the top of her tail. This "straddle walk" is a common mating behavior in terrestrial salamanders.

Male spotted salamanders migrate from underground terrestrial retreats to ephemeral pools in late winter. They deposit spermatophores (*right*) on the substrate that females pick up and use to fertilize their eggs internally. Following internal fertilization, female spotted salamanders attach their eggs (*below*) to twigs and grass stems in shallow water. The eggs will develop for several weeks before hatching.

Courtship behavior in the lungless salamander group (the Plethodontidae) is remarkably uniform among species. The following sequence is typical whether it occurs in water or on land: (1) the male orients to and approaches a female; (2) the male makes head contact with female; (3) the pair engages in a "tail-straddling walk" in which the female places her chin on the base of the male's tail and positions herself over his tail; (4) the male deposits a spermatophore in front of her snout; (5) the pair moves forward and the female lowers her cloaca over the spermatophore and picks up the sperm packet with her cloacal lips. Fertilization is internal. Males in this group have a remarkable way of enticing a female to be receptive that is called "vaccination." During the mating season, the male grows large teeth in the front of his upper jaw, and the mental gland under his lower jaw grows and enlarges. During the elaborate courtship, the male uses his enlarged front teeth to scratch the skin of

DID YOU KNOW?

Many salamanders use tail waving to distribute chemicals known as pheromones to attract mates. An uplifted tail can also distract potential predators away from the salamander's head and body.

top Hatchling spotted salamanders have gills and all four limbs. These aquatic larvae will grow and develop in the water until they metamorphose into juveniles.

bottom The metamorphs are small and often have remnant gill buds behind the head. The juveniles will live in the terrestrial environment for another 2–3 years before they mature and return to the pond to mate.

the female. He then presses his enlarged mental gland against the abraded skin and releases a secretion from it that enters the female's body and presumably makes her receptive for mating. Females then lay their eggs under logs or rocks or in depressions in the soil and guard them from predators until they hatch. Females of some species produce antibacterial, antifungal skin mucus that protects the developing eggs. Eggs of aquatic plethodontid salamanders hatch into gilled larvae; all of the larval development of terrestrial salamanders occurs within the egg, and the hatchling looks like a miniature adult.

Newts have a complex courtship ritual that is easily observed in an aquarium. The male rubs his forearm across the female's snout, then presses his cheek against her snout as he lifts his front leg under her chin. He grasps her around the neck in front of her front limbs, fans his tail back and forth, jerks his tail base, and thrashes his body while continuing to rub her snout with his cheek, all the while dragging the female roughly about. After several minutes he dismounts, moves in front of the female, and while moving forward wiggles his body and arches his tail. The female nudges his cloacal region or his tail, and the male deposits a spermatophore on the bottom. He then moves forward a body length and turns sideways to block her path. She positions herself over the spermatophore and picks up the sperm packet with her cloaca. During the entire process the male is often harassed by other males attempting to dislodge him, although they usually are not successful. The eggs are fertilized internally, and the female deposits each egg singly on a leaf blade and folds it over, gluing it shut with mucus from her cloaca. Egg laying may take several days. Newts show no parental care, and the aquatic larvae hatch in 3–4 weeks.

Age and Longevity People often equate small body size with short lives and large body size with long lives, but salamanders are surprisingly long-lived for their relatively small size. Red-spotted newts, for example, metamorphose

The jelly surrounding spotted salamander eggs takes up water and swells to form a protective gelatinous mass that may be clear or opaque.

from the larval stage to the eft (juvenile) stage several months after egg laying. The terrestrial efts then forage in the forest for 2–7 years before they mature and return to their home pond to reproduce for the first time. Adults often live for another decade after that. Longevity in natural populations is largely unknown, but in a long-term South Carolina study, individually marked mole salamanders were recaptured more than a decade after they were marked.

Most of the available records on longevity are from long-term captives. Small salamanders such as Blue Ridge two-lined, northern dusky, and northern slimy salamanders have been kept in captivity for as long as 5 years. Larger salamanders such as amphiumas and spotted salamanders are known to live more than 25 years. Of all eastern salamanders, the hellbender has been reported to be the longest-lived at more than 30 years. Age at maturity for the small salamanders is often 1–2 years but is as long as 5–10 years for the largest salamanders. Some deer, rabbits, and songbirds, in contrast, reach maturity within a year or just a few months of birth. Many salamanders, that is, live much longer than almost all game species. Among other things, this means that salamander populations are very slow to recover when reproductive adults are lost as a result of habitat de-

The age of salamanders can be determined only by marking them as first-year hatchlings or metamorphs or by taking a slice out of a leg bone and analyzing the number of growth rings under a microscope.

struction or alteration, even if the habitat rebounds to its natural state. Think about the former natural places you know that have been changed in some way, such as land where the hardwood trees were removed through timber harvesting. We should not be surprised that salamanders throughout the eastern United States are declining in numbers. Specific conservation issues are noted in each of the species accounts and are summarized later in this book.

SALAMANDER DIVERSITY AROUND THE WORLD

Most of the world's 816 recognized species of salamanders are found in the Northern Hemisphere; the group is entirely absent from the continents of Antarctica, Australia, and most of Africa. The only tropical salamanders are in the family Plethodontidae, and they occur mostly in the mountains of Central America and the Amazon basin. These ancient amphibians have adaptations that allow them to occupy many types of freshwater and terrestrial habitats. No salamanders anywhere in the world can tolerate salt water; thus no salamander species are found in oceans, salt marshes, or tidal creeks. Species of some salamanders are found on barrier islands of the Atlantic and Gulf coasts, but they are restricted to freshwater habitats or to woodland habitats not regularly flooded by saltwater.

Many salamander species are found only in high mountains, especially in the Appalachians and in Central America. The Siberian newt (*Salamandrella keyserlingii*) occurs as far north as 64° north latitude in the Russian Plain—almost to the Arctic Circle. This amazing salamander has the largest range of any species in the world and is capable of surviving after being frozen in the tundra ice for several years. One such individual excavated from a depth of 33 feet was estimated to be 90 years old! Although the Siberian newt is the extreme case, many salamanders can tolerate subfreezing temperatures for varying lengths of time because they have an antifreeze compound called glycogen in their tissues and organs. Salamanders tend to be more active at and more tolerant of cooler

DID YOU KNOW?

No salamanders are able to tolerate salt water or brackish water; thus none occur in coastal marshes or estuaries.

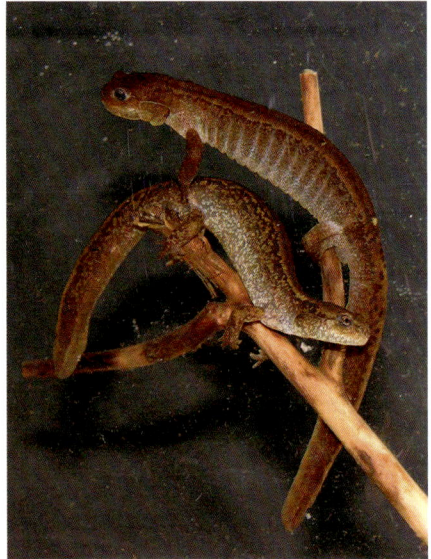

The Siberian newt (*Salamandrella keyserlingii*) is found as far north as 64° north latitude in the Russian Plain—almost to the Arctic Circle. It can withstand long periods of being completely frozen.

European fire salamanders (*Salamandra salamandra*) from Eastern Europe exhibit phenomenal variation in colors and patterns. These salamanders got their common name when they were observed emerging from logs put on fires.

temperatures than frogs, with the exception of the wood frog (*Rana [Lithobates] sylvatica*). With more than 115 species, 17 genera, and 7 families, the Southeast indisputably has the highest biodiversity of salamanders in the United States or most comparably sized regions in the world.

Three regions in North America outside the eastern United States contain high diversities of salamanders as well: the mountains of the Pacific Coast; the Edwards Plateau region in central Texas; and the Ozark and Ouachita mountains in Arkansas, Missouri, and Oklahoma. Two families—Dicamptodontidae with 4 species in the genus *Dicamptodon*, and Rhyacotritonidae with 4 species in the genus *Rhyacotriton*—occur exclusively in the mountains and surrounding area of the Pacific Coast, along with 3 ambystomatids, 46 plethodontids, and 4 newts. Thirteen species of plethodontids, all in the genus *Eurycea*, occur only in the streams and caves of the Edwards Plateau. Ten members of three genera (*Ambystoma*, *Eurycea*, and *Plethodon*) occur primarily in the Ozark region. One is the bizarre grotto salamander (*Eurycea spelaea*), which has adapted to life in caves. Adults are white to pinkish white with black spots for eyes and spindly

above Members of the tropical genus *Oedipina*, such as this endangered *Oedipina poelzi* from Costa Rica, are characterized by an exceptionally long tail.

right The Mexican axolotyl (*Ambystoma mexicanum*) lives in lakes around Mexico City. These salamanders never metamorphose and reach enormous sizes.

The Mexican mushroom salamander (*Bolitoglossa mexicana*) is a high-elevation salamander found in Central America.

left Members of the genus *Thorius* are the smallest salamanders on earth. These tiny salamanders occur on isolated mountaintops in Mexico.

right The bizarre grotto salamander (*Eurycea spaeleus*) of Texas has completely adapted to life in caves. Adults are white to pinkish white with black spots for eyes and spindly legs.

above European fire salamanders (*Salamandra salamandra*) can squirt their poison at predators.

right The Japanese giant salamander (*Andrias japonicas*) is closely related to the North American hellbender. It lives in streams and reaches a total length of nearly 5 feet. This individual was 4 feet 9 inches long.

legs. All seven currently described sirens (family Sirenidae) are confined to midwestern and southeastern North America. All three members of the Amphiumidae (*Amphiuma*) are found in the Southeast. One or two members (depending on which taxonomy is accepted) of the family Cryptobranchidae occur in North America (hellbenders), and Asia has three species of giant salamanders in China and Japan.

Only one family occurs exclusively outside North America. The Asiatic salamanders in the family Hynobiidae inhabit Asia and most of Russia. Most are less than 4 inches in total length, but one, *Ranodon sibericus*, reaches almost 10 inches. Except for the European olm (*Proteus*, in the family Proteidae), all seven mudpuppies and waterdogs are confined to North America. The newt family (Salamandridae) has the largest global distribution of any group of salamanders. Several species are found in Europe, Asia, and eastern and western North America. Many of the newts are brightly colored, and most are quite toxic as a result of the poison secreted by their abundant skin glands. Several, such as the European fire salamander (*Salamandra salamandra*), actually squirt their poison from glands at predators, making them perhaps the only venomous amphibians. Two European species of newts bear their young alive instead of laying eggs as all other salamanders do.

The evolutionary process has produced myriad body shapes and sizes, but almost all salamanders have a similar appearance: elongate body, long tail, and relatively small legs compared with other terrestrial vertebrates. Salamanders worldwide have a body shape ranging from slender to robust, a distinct head, and four limbs, and the full range of aquatic and terrestrial body forms are found in the eastern United States. Only the amphiumas, the sirens, and the European olm have an elongated, eel-like body with much reduced or lost limbs.

The smallest eastern salamanders are the pygmy salamanders (*Desmognathus wrighti* and *D. organi*) and the patch-nosed salamander (*Urspelerpes brucei*), none of which reach a maximum total length of 2 inches. The largest species worldwide is the endangered Chinese giant salamander (*Andrias davidianus*) in east-central China, which may reach 6 feet in total length. The Japanese giant salamander (*Andrias japonicas*) is a close second. The related hellbender closely resembles these giants in body form and habits.

SALAMANDER TAXONOMY

Taxonomy is the scientific field that deals with the classification and naming of plants, animals, and other organisms. Scientists engaged in this field classify and name salamanders in ways that reflect ancestral genetic relationships among species. Closely related species are placed within the same genus (plural = genera), and closely related genera are placed within the same family. All described organisms have a two-part scientific name, which is italicized to reflect its Latin or Greek origin. The genus name (e.g., *Plethodon*) is always capitalized. The second, or species, name (e.g., *cinereus*) is not. The example given here is the scientific name of the red-backed salamander, *Plethodon cinereus*. The name of the northern slimy salamander is *Plethodon glutinosus*, and that of the Shenandoah salamander is *Plethodon shenandoah*. The inclusion of these three species in the same genus indicates that they are more closely related than other species that are in the same family but in a different genus, such as *Eurycea* or *Pseudotriton*. Scientific names can be abbreviated after first usage to avoid repeating the genus (e.g., *P. cinereus*) as long as no ambiguity results.

Taxonomy is not a static science. Early descriptions of new salamander species were based on the scientist's knowledge of the diversity of these animals in the region and recognition that the specimen in question was different enough to warrant a new name. Many eastern salamanders were described in the 1800s and early 1900s in just this way. Molecular biology has revolutionized our understanding of the genealogies/phylogenies of species, genera, and families. Several techniques allow taxonomists to examine the genetic structure of species, leading to a much more refined understanding of the relationships of populations in different areas. These new techniques combined with improved accessibility to remote regions have resulted in descriptions of many new salamander species. The description process continues today and will continue as new species are described. The species accounts included here indicate which of the currently recognized species may actually include one or more new species and when each species was first described.

Modern genetic techniques have also produced questions about how "species" should be defined. Historically, a species has been defined as a group of interacting individuals and populations that can reproduce and leave viable offspring. Two or more distinct species living in the same place that do not hybridize are not able to produce offspring and are thus said to be "reproductively isolated." Groups of populations that we once thought were not isolated, however, may in fact be different enough to warrant recognition at the species level. How different do populations have to be to deserve specific recognition? Scientists are still debating that question.

Some salamander species described on the basis of genetic differences are so similar in appearance to another species that they cannot be differentiated morphologically; only molecular

The eastern red-backed salamander (top), the northern slimy salamander (middle), and the Shenandoah salamander (bottom) are all members of the genus Plethodon.

techniques and/or the geographic location can identify them. Examples include the three species of gray-cheeked salamanders in the Blue Ridge Mountains of North Carolina, the Valley and Ridge salamander and the Shenandoah Mountain salamander in western Virginia, and the mountain dusky salamander group in the southern Appalachians. Fortunately, most salamanders have enough observable differences to distinguish them in the field.

Common, or standard, English names are not based on a set of rigid protocols as scientific names are. The common name of a species, particularly a widespread one, is likely to vary regionally. The spelling of a name, whether two words are hyphenated, and other details are hotly debated in some circles, although the differences are often very minor. Many people would like to see common names standardized as scientific names are, and so the debates continue. This book generally uses the scientific and standard English names in the most recent version of Amphibiaweb (https://amphibiaweb.org/), which is accepted by most professional herpetologists.

Northern (*top*) and southern (*bottom*) gray-cheeked salamanders are closely related and are so similar in outward appearance that the only reliable way to differentiate between them is at the molecular level.

HOW TO IDENTIFY SALAMANDERS

Visually identifying and differentiating between most species of eastern sala-manders is relatively straightforward on the basis of body size, shape, color, and pattern. The fact that numerous species of terrestrial and stream-breeding salamanders have very well defined geographic ranges that do not overlap sim-plifies identification in some situations because all or most of the species that look similar can be eliminated from consideration on the basis of locality. Pop-ulations of some species show remarkable variation in color and pattern, and learning the range of variation in the salamanders in your area may be necessary to ensure accurate identification. For the majority of salamanders, however, a few particular characteristics can be used to determine the family and genus, and in many instances the species. Several basic features of appearance are commonly used in identifying most species of eastern salamanders and differ-entiating them from other species that might be in the vicinity.

Body Size Body size can be measured as the length of the body (snout to vent [the posterior margin of the cloacal opening]), as the body plus tail length, or as the weight of the body (mass). The maximum size of adults of a species is sometimes a useful character in differentiating among species that are similar in appearance, such as the dwarf and two-lined salamanders. Minimum and maximum lengths in this book are given as total length (body plus tail), so when you are using size to identify a species, it is important to consider whether an individual's tail has been broken. The adults of some salamanders are sexually dimorphic; usually the female is the larger sex.

Body Shape All salamanders have a basic elongated shape, but species can be characterized as slender or stout, and some have a flattened head and body. The proportional tail length can also be a distinctive feature; the long-tailed salamander, for example, has a tail that is nearly two-thirds of the total body length. The shape of the tail varies among species from the oarlike tail of fully aquatic sirens and mudpuppies to the tails of some dusky salamanders that are round in cross section at the base. Once you are familiar with all the sal-amanders in your area, you will find that identification can often be made on the basis of shape alone.

Legs and Toes The absence of hind legs immediately differentiates the sirens and dwarf sirens from all other salamanders. The remaining species of sala-manders all have four legs, but in proportion to their body size. The tiny legs of amphiumas are distinctive enough to keep any species in this genus from being confused with any other group of salamanders. The number of toes on front or

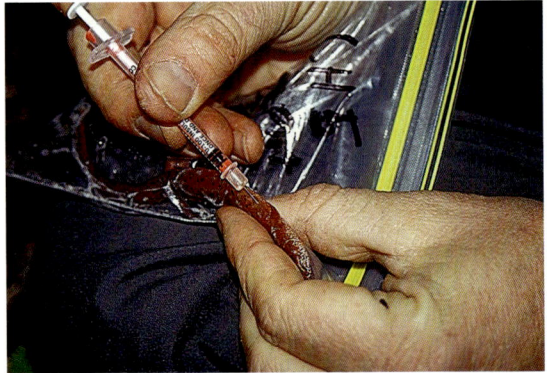

Salamander biologists sometimes draw blood for use in genetic analyses to determine relationships among salamanders and to identify some species.

The absence of rear legs immediately distinguishes the sirens and dwarf sirens (such as the individual shown here) from all other salamanders, including the similar amphiumas.

hind feet can also be distinctive; the amphiumas, for example, have one, two, or three toes per foot depending on the species. Four toes on each back foot rather than the five present in most salamanders are characteristic of the dwarf salamanders, four-toed salamander, waterdogs, and mudpuppies.

Costal Grooves Vertical grooves, which are visible along the sides of most salamanders, indicate the number of vertebrae and are an effective way to identify some species. For example, the mole salamander has only 10 costal grooves, many of the dusky salamanders have 14, and greater sirens have as many as 40. However, newts have no costal grooves.

Genetics Many species of salamanders, especially within the family Plethodontidae, are "electromorphs" or cryptic species; that is, they differ genetically but are identical in appearance. Even salamander biologists cannot distinguish between some species without using genetic analysis techniques. The exact location of capture can help in determining the identity of most species that look similar, but not all. Three factors make identification difficult in such cases. First, many closely related species interbreed in their zones of overlap, resulting in hybrids that have characteristics of both species. The occurrence of hybridization among various forms is noted in the species accounts. Second, two or more salamander species may be virtually identical in body shape, color pattern, habitat, and behavior, but because they differ in their genetic makeup they qualify as separate species. Thus, an individual salamander belonging to one of the 10 species of slimy salamanders that occur in the Southeast may not differ appreciably in appearance from Virginia to Florida to Louisiana. Third, individuals of some species exhibit a range of body colors and patterns within the same population. The Ocoee salamander, for example, may be nearly black with a chestnut spot pattern on the back or may be orange without distinct markings—or may exhibit some variation of these characters. The best approach is to learn how the populations and species in your area vary in color and pattern.

Body Markings Particular colors and patterns on the head, body, legs, and tail are characteristic of many species of salamanders. The presence, position, width, shape, and intensity of dark or light stripes down the back and onto the tail are key features in the identification of two-lined, three-lined, dwarf, and zigzag salamanders; striped newts; and others. The size and pattern of dark or light spots can often be used to distinguish between mud salamanders and red salamanders or between slimy salamanders and some of the other woodland species. Individuals within the same population of eastern or southern red-backed salamanders may have a prominent reddish stripe down the back or no stripe at all. The two sexes of the marbled salamander differ subtly in their black and white coloration, with the males being a brighter white.

Special Head Characters Several head, jaw, and chin features are characteristic of some or all species in the family Plethodontidae and may even classify an individual salamander as belonging to a particular genus.

NASOLABIAL GROOVE The presence of a thin groove that runs from the nostril to the upper lip and is visible to the naked eye is a signature trait of members

The vertical indentations (costal grooves) between the legs of many salamanders correspond with the number of vertebrae in the backbone and are used in some cases to distinguish between species.

of the family Plethodontidae. Although not useful for distinguishing among species within the family, the presence or absence of the nasolabial groove can readily be used to separate plethodontid salamanders from other families of salamanders such as mole salamanders and newts.

CIRRI Some species, such as the southern two-lined salamander (genus Eurycea), have a pair of fleshy projections that point downward from the upper lip. Cirri are most prevalent in males during the breeding season but are present in females of some species.

MENTAL GLAND An enlarged round or oval disk called the mental gland is present beneath the chin of males in the genus Plethodon during the breeding season. The mental gland is more prevalent in some Plethodon species than others. It is absent in salamanders in other genera, with the exception of the pygmy salamander, which also has a mental gland.

CANTHUS ROSTRALIS The spring salamander has a distinctive bony ridge known as the canthus rostralis that runs from the eye to the nostril. The ridge is usually marked by a dark-bordered white line.

EYE–JAW LINE Dusky salamanders can usually be distinguished from other species in their range by an obvious light line that runs posteriorly from the eye to the jaw.

Downward projections (cirri) in many of the male plethodontid salamanders, such as this three-lined salamander, appear in the breeding season and are likely used for stimulation of and communication with females.

During the mating period the mental (or hedonic) gland on the chin of male plethodontids, such as this Chattahoochee slimy salamander, enlarges.

left Spring salamanders have a distinct ridge bordered by a light line along each side of the snout between the eye and nostril.

below All members of the genus *Desmognathus* have a light line—the eye–jaw line—that runs from the eye to the back of the jaw.

TEETH Tooth shape is important in distinguishing some closely related species. For example, the northern dusky salamander has narrow crowns on its teeth, whereas its close relative the flat-headed salamander has relatively wider crowns.

Skin Texture Most salamanders have soft skin that can range from the rubbery feel of the mole salamanders to the mucus-covered bodies of sirens, mudpuppies, and amphiumas. Newts in the terrestrial eft stage have granular skin that differentiates them from other forest-dwelling salamanders in the eastern United States.

External Gills More than 70 species of eastern salamanders are permanently aquatic or have aquatic larvae. Most of them have external gills that are visible outside the body, although adult amphiumas and hellbenders have gill openings but no visible gills. The more than 45 woodland species that are terrestrial breeders bypass the typical larval stage

DID YOU KNOW?

Some salamanders have lungs, some have gills, and some have both lungs and gills, but all can respire through their skin. Many of the woodland salamanders have no lungs or gills and respire only through their skin.

The external gills of Tennessee cave salamanders are used for respiration. The rows of light dots on the head and body of this individual are part of the lateral line system used to detect movement through water.

and develop completely within the egg. The hatchlings have no external gills, although gill buds—remnants of the embryonic gills—are sometimes present.

Environmental Information The environmental conditions under which a salamander was found can often help in identifying it. For example, knowing that marbled salamanders emerge from underground retreats and breed in the first rains of the fall will help to distinguish them from tiger or spotted salamanders, which breed months later.

Geographic Location The geographic location where a salamander is found can sometimes provide a clue to its identity by eliminating similar species not found there. A waterdog from a river in southern Alabama, for example, can be identified as a Gulf Coast waterdog simply on the basis of where it was found.

Habitat The specific habitat, such as aquatic or terrestrial, stream or pond, forest or clearcut, can be very helpful in species identification. The species accounts indicate habitats where species are most likely to be found.

Time of Day and Year Some eastern salamanders migrate during their breeding season and are more likely to be found in some areas than in others. In the case of some woodland salamanders, the movement is vertical, from the surface to deeper underground. Webster's salamanders are highly unlikely to be discovered beneath a log or rock in July or August when they have retreated underground, although they may be common at such sites in January or February. Different species of seasonal wetland–breeding salamanders can be expected to be moving at night during rains at specific times of the year. With the exception of newts, which are occasionally seen moving about during the day, salamanders characteristically move about above ground on humid nights.

WHY ARE THERE SO MANY NEWLY DESCRIBED SALAMANDERS?

The number of described salamander species has steadily increased over the last 40 years. This increase is the result of advances in molecular technology, ranging from electrophoretic analysis of isoenzymes (early on) to sequencing of mitochondrial DNA and of whole genomes or, most recently, parts thereof.

Many salamander species that were once considered to be a single, widespread species have been found to comprise groups of cryptic species that have diverged in terms of their mitochondrial and/or nuclear DNA. These salamanders tend to be both ecologically and morphologically conservative (i.e., physically identical), which means that speciation has often occurred without accompanying morphological and ecological differences.

Recent research on salamanders, especially the lungless salamanders (Family Plethodontidae), has revealed that many species long considered valid single species are, in fact, groups of cryptic species. It has often been assumed that these cryptic species are closely related, but that, too, has proven not to always be the case.

The new molecular technologies have the potential to greatly change our understanding of salamander biodiversity. We can, therefore, expect dozens of new genetically distinct species to be described in the coming years. Nevertheless, as molecular biology advances and new species are described, increased scrutiny as to the validity of these proposed new species is warranted.

A major concern with the accelerating rate of new species descriptions is that an understanding of their natural history, ecology, and conservation lags behind their discovery and description. Like most species on earth, amphibian populations are declining rapidly. Most of the newly described species currently do not have legal protection, even though some are restricted in distribution and habitat. Data on the biology and ecology of these newly described species are urgently needed so that conservationists at state and federal levels can protect and properly manage our treasured salamanders.

SALAMANDER HABITATS IN THE EASTERN UNITED STATES

Eastern salamanders live in a variety of aquatic and terrestrial habitats. Some species are strictly aquatic and never go on land; others are strictly terrestrial. Many species require both types of habitat in order to complete their life cycle. Species' adaptations to one or more of these habitat types have led to natural partitioning into ecological groupings that generally, although not always, follow family lines.

All eastern species in the families Amphiumidae (3 species), Cryptobranchidae (1 species), Proteidae (7 species), and Sirenidae (7 species) are permanently aquatic and seldom if ever venture onto land. Six of the 90 eastern species in the family Plethodontidae are aquatic as well, while 48 species are terrestrial in all stages of their life cycle and are found exclusively in hardwood, pine, mixed hardwood, or spruce-fir mountain forests. All 10 of the eastern species of Ambystomatidae species, 2 species of newts (family Salamandridae), and the other 42 species of plethodontids live in or around forest habitats but breed in aquatic areas. Some lay their eggs in small streams or seepage areas, others in isolated wetlands.

Some habitats have a greater abundance and diversity of species than others. Because salamanders are so seldom seen, however, the high number of species and individuals may go unnoticed by anyone but a herpetologist. Knowing the general habitat types favored by salamander species as well as the microhabitats where they are most likely to be found is the first step toward enjoying a group of fascinating creatures that few people ever see.

DID YOU KNOW?

As most amphibians spend some of their life on land (terrestrial) and some in the water (aquatic), among all animals they are one of the best biological indicators of an ecosystem's health.

opposite Dusky salamanders live in rocky streams in mountainous regions.

right Georgia blind salamanders are completely aquatic and are found in limestone caves, sinkholes, and artisan springs.

Salamander Associations with Freshwater Wetland Habitats The aquatic habitats of the eastern United States can be categorized on the basis of several characteristics. Seasonal variation in size and permanence, moving/flowing water or standing water, annual extremes and variation in temperature and rainfall, water chemistry, and the presence or absence of other species—especially fish—are all factors determining which salamanders will be found in a body of water. The following generalized categories pertain to identifiable aquatic habitats with which various salamander species are associated.

PONDS, LAKES, AND RESERVOIRS Permanent lakes and ponds, such as those formed by beavers damming small streams or those created where the earth collapses in some limestone areas to form sinkholes, are natural habitats. Man-made farm ponds, which may function much like beaver ponds, are also liberally distributed throughout the region. Large, deep, natural lakes of glacial origin are present in northern regions, and artificial reservoirs created by damming large rivers are physically similar. A few salamander species are associated with these habitats, but except for the mudpuppy, which can survive in large lakes where fish are present, the eggs, larvae, or even the adults of most end up as prey. Salamanders lay their eggs along the flooded peripheries of permanent ponds, lakes, and reservoirs where emergent vegetation is present and the water is shallow enough to preclude the presence of predatory fish.

ISOLATED SEASONAL WETLANDS Isolated seasonal wetlands are found throughout the eastern United States. They come in different shapes and sizes

Natural ponds such as this Carolina bay support populations of mole salamanders, sirens, amphiumas, and newts.

left Vernal ponds are seasonal wetlands that often dry during late summer and are fish-free during the period when many of the mole salamanders breed.

right Tiger salamanders and other mole salamanders have been observed breeding in ponds with ice on the surface.

Sinkhole ponds and isolated ponds created by humans are used extensively by pond-breeding salamanders and newts.

left Bogs with shallow water and soft, mucky substrates are usually spring fed and support four-toed salamanders, mud salamanders, and red-spotted newts.

right Mudpuppies and hellbenders are found in cold, shallow rocky streams.

and have several geological origins. The Carolina bays common in much of the Coastal Plain from southeastern Georgia to Virginia, vernal ponds, glacial potholes, sinkholes, lime sinks, and scoured depressions in stream or river floodplains have basic features that make them ideal for certain groups of salamanders. Many dry up in some years, and some dry up every year, so they are unlikely to have fish that prey on salamanders or their eggs. Some species of mole salamanders such as the marbled and spotted salamanders require isolated fish-free wetlands for breeding success. Salamanders can breed in seasonal wetlands that vary in surface area from smaller than a backyard to several acres. The depth may also vary, from a few inches in some to more than 10 feet in others. Among the factors determining how long an isolated wetland holds water during a year are the porosity of the substrate (e.g., some Carolina bays have a clay lens that helps retain standing water), the amount and timing of rainfall, evaporation rates (which in turn depend on air temperature and humidity), transpiration rates (which depend on the amount and type of standing vegetation and how much water is released through the leaves to the atmosphere), and proximity to streams or rivers that might flood. Regions of the eastern United States where isolated wetlands are scattered abundantly across the landscape are home to numerous forest-dwelling salamander species that breed aquatically, including mole salamanders, newts, and dwarf salamanders.

BOGS Mountain bogs in the Blue Ridge Mountains, pitcher plant bogs in the Coastal Plain, and wetlands where small streams meander through old pastures are usually small, acidic, and fish-free. Newts, four-toed salamanders, and mud salamanders sometimes breed in these wetland habitats. One-toed amphiumas can be found in pitcher plant bogs.

RIVERS, LARGE STREAMS, AND SPRING RUNS The rivers and large streams of the eastern United States and the clear limestone springs and streams of Georgia and Florida are impressive aquatic habitats that set the cultural character and often shape the natural biological features of a region. But aside from marginal areas of flooding such as river swamp bottomlands, they are not home to many salamanders. Hellbenders inhabit mountain streams with large rocks and riffles. Sirens and amphiumas, which reach sizes large enough to deter most fish predators, are sometimes found in Coastal Plain rivers and streams. The 6 species of permanently aquatic waterdogs inhabit streams and rivers from Virginia to Louisiana in different parts of the Coastal Plain and Piedmont.

Small, rocky streams throughout the Southeast are habitat for a variety of small salamanders with an aquatic larval stage.

Two-lined and three-lined salamanders, red and mud salamanders, and many of the dusky salamanders thrive in stream or seepage areas within hardwood forests.

FLOODPLAIN RIVER SWAMPS, OXBOW LAKES, AND BOTTOMLAND SWAMPS The seasonal drying and flooding regimes of floodplain river swamps, backwater oxbow lakes, and bottomland swamps are particularly dependent on local and upstream rainfall. The extensive floodplain swamps that border the slow-moving rivers and large streams of the southeastern Coastal Plain are breeding sites for newts and several species of forest-dwelling mole salamanders. Plethodontid species such as slimy salamanders, three-lined salamanders, and red salamanders may be present in the surrounding swamp forests.

SMALL STREAMS AND SEEPAGE AREAS Small, rocky streams formed from mountain springs and watershed runoff are habitats with high salamander diversity and abundance, as are the wetland habitats known as seeps. Seeps can be created by groundwater slowly flowing out at the base of a steep bluff or between bedrock formations. Little rivulets less than a foot wide and only a few inches deep may flow away from a bluff, sometimes accumulating into small streams that flow into larger ones. Leaf litter and debris are common characteristics of such sites. Dusky, two-lined, dwarf, mud, and red salamanders are often abundant in seeps and may be found hiding in and under the debris. Rocky streams flowing through forest habitat, even small ones whose flow is noticeable mostly during wet periods, may still have high numbers of stream salamanders such as black-bellied, spring, dusky, and Junaluska salamanders.

Salamander Associations with Terrestrial Habitats The terrestrial habitats of the eastern United States can be categorized according to their vegetation cover, soil type, topography, and geological history. The following generalized categories describe identifiable terrestrial habitats with which various salamander species are associated.

MIXED HARDWOOD FORESTS Hardwood forests, which occur throughout the eastern United States, can include or may even be dominated by a broad array of such tree types as oaks, hickories, magnolias, maples, and gums. Many mixed hardwood forests have interspersed pine trees. In their natural state, these forests offer vital hiding places for salamanders beneath the ground litter of decaying leaves, dead bark, and logs; and their thick canopy provides shade and helps to retain moisture. More than two-thirds of eastern salamander species can be found in mixed hardwood forests. Adult mole salamanders and the eft stage of newts spend most of their lives in such forests, but always within migrating distance of wetland breeding habitats. Two-lined salamanders, red salamanders, and many of the dusky salamanders depend on healthy hardwood forests being adjacent to the stream or seepage areas where they breed and spend much of their time. Most woodland salamanders (family Plethodontidae)

Although they are amphibians, some salamanders are strictly terrestrial and do not require ponds or other wetlands for survival. Logs and other surface objects in terrestrial habitats are used extensively by several salamanders.

Many forests in the eastern United States are composed of a mix of hardwood and pine. Fewer species occur here due to the drier substrate.

Although longleaf pine forests and flatwoods occur on sandy and xeric soils, several southeastern salamanders are able to find shelter in these upland habitats.

are terrestrial breeders that live their entire lives in mixed hardwood forests, sometimes on the surface and sometimes in burrows, crevices, or other deep underground retreats.

PINE FORESTS Several species of pine trees can be found scattered among hardwood forests at all elevations and under differing soil and moisture conditions, but pine forests abound in the Atlantic and Gulf Coastal Plain from New Jersey to Texas. Most of our longleaf pine–wiregrass forests are mere remnants of the once vast fire-maintained savannas that covered much of the southeastern Coastal Plain. Even the dry longleaf pine forests of the sandhills can support a variety of forest-dwelling salamanders such as tiger salamanders, which survive in underground burrows and travel to seasonal wetlands for breeding. The low-lying pine flatwoods habitat found closer to the coast once dominated large areas of the Southeast. Pine flatwoods, which consist primarily of longleaf and slash pines, once made up almost half of Florida's natural habitats. Seasonal wetlands in these habitats serve as breeding sites during the winter for striped newts and federally protected flatwoods salamanders.

SPRUCE-FIR FORESTS The steep, rocky, mountainous terrain at higher elevations in the Appalachian Mountains of West Virginia, Pennsylvania, New York, Vermont, New Hampshire, Virginia, North Carolina, and Tennessee, including Great Smoky Mountains National Park, is the home of coniferous forests dominated by spruce and Frasier fir. The heavy layer of moss coupled with the extensive rocky latticework of underground hiding places and continually moist conditions form ideal habitats for terrestrial-breeding species such as red-legged and pygmy salamanders.

above A stream flowing through a cave can be an ideal habitat for several species of salamanders, some of which may be permanently aquatic.

right Some woodland salamanders are associated with limestone rock outcrops and disappear deep underground during hot, dry weather.

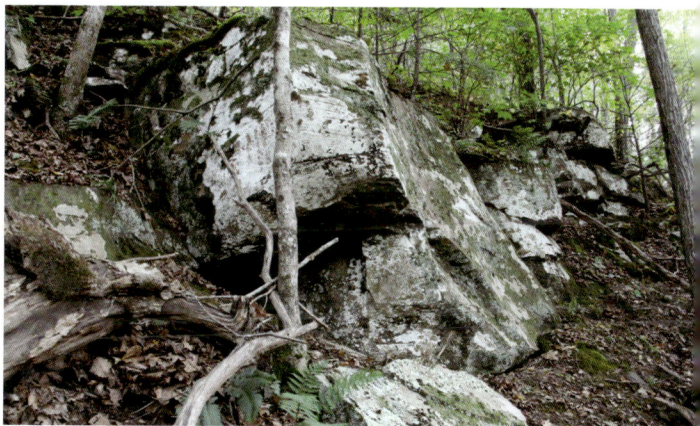

LIMESTONE CAVES Common features of the landscape in areas of karst topography are caves and crevices created as the limestone dissolved away over the eons. In addition to the cave salamander, several forest-dwelling species—including Pigeon Mountain, Cumberland Plateau, slimy, long-tailed, and zigzag salamanders—will enter caves, sometimes for weeks or months at a time during periods of drought or extreme cold or heat. The Georgia blind salamander, West Virginia

DID YOU KNOW?

Numerous salamanders in the eastern United States spend part of their lives deep in caves. Some remain in the twilight zone, but many go farther inside to overwinter and escape predators on the surface.

Agricultural landscapes are usually inhospitable to salamanders, but these amphibians will cross long distances to reach wetlands for breeding.

spring salamander, Tennessee cave salamander, and Berry Cave salamander are found only in caves and live permanently in pools associated with underground streams and rivers.

URBAN, SUBURBAN, AND AGRICULTURAL HABITATS Salamanders differ from other major groups of vertebrates—including frogs, snakes, lizards, and turtles—in being unable to thrive as well in areas developed by humans as they do in natural areas. A few species such as red-backed and slimy salamanders can persist indefinitely in urban parks, suburban areas where small, wooded tracts are left intact, and even on golf courses if fish-free wetlands are present. Tiger salamanders sometimes burrow in agricultural fields, but they must also travel to aquatic breeding sites and therefore may have to survive long-distance migrations that involve crossing busy highways. Two-lined salamanders and many of the dusky salamanders can tolerate degraded streams in urban zones.

ORGANIZATION AND ORDER OF SPECIES ACCOUNTS

The names used in the species accounts—both the scientific name and the common name—are those generally accepted by salamander biologists in 2022. The Taxonomy section in each account lists alternate names that have been used or proposed for the species. Each species account includes a description of adults, larvae, and juveniles, as well as other salamander species that are similar in appearance. The accounts describe habitat, behavior, activity patterns, reproduction, diet, and predators. Each account also includes a discussion of relevant conservation issues.

In contrast to other amphibians (frogs) and reptiles (crocodilians, snakes, turtles, lizards) of the eastern United States, a high proportion of our salamanders have highly localized distribution patterns, sometimes being known from only a single mountain in one or a few counties. The regional distribution in the Southeast or East, as well as the U.S. distribution of certain species, is presented on maps that accompany each account.

The accounts are grouped into three major assemblages: aquatic, semiaquatic, and terrestrial. They are further partitioned into categories based on salamander life-cycle patterns and the habitats with which species are most likely to be associated. Some salamander species retain their larval gills and remain aquatic their entire lives. Many salamanders, such as the stream salamanders, are semi-

Frosted flatwoods salamander populations have declined dramatically in the past four decades. This species is on the Endangered Species List maintained by the U.S. Fish and Wildlife Service.

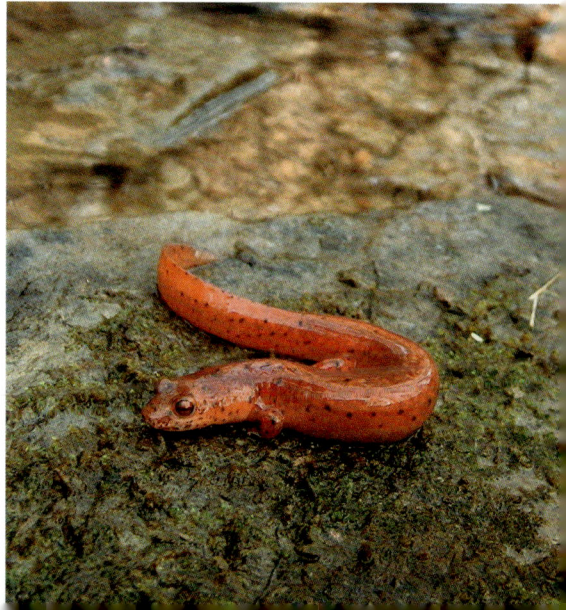

Spring salamanders are the largest plethodonid salamander species in mountain streams, where they prey on other salamanders.

aquatic; that is, they breed in water but spend approximately 11 months of the year in the surrounding forest. The woodland salamanders (genus *Plethodon*) do not require water at all for breeding and larval development. They lay their eggs in moist places under rocks or in logs, and development takes place inside the egg capsule. The metamorph (i.e., hatchling) of such species is a miniature of the adult body form. These terrestrial-breeding woodland salamanders are thus able to exploit the moist, shaded environment of hardwood forests. *To see the order of the species accounts at a glance, consult the table on pages 462–465.*

Aquatic Salamanders Eighteen species of eastern salamanders rarely or never move about on land. Most live in rivers, streams, swamps, lakes, or ponds, although some also occur in seasonal wetlands that do not have year-round permanent water. The diversity of these fully aquatic salamanders is greater in the Southeast than anywhere else in the world.

BASIC FEATURES OF THE SPECIES ACCOUNTS

Common and scientific names

WEBSTER'S SALAMANDER *Plethodon websteri*

Quick identification guide

BODY SHAPE	BODY COLOR	PATTERN	BODY SIZE
Small and slender	Brown or reddish brown	Usually a broad stripe down the back and onto the tail	Adults up to 3.3 inches in total length
NUMBER OF COSTAL GROOVES 17–19			

Descriptions of adults, larvae, and juveniles of the species

Description Webster's salamanders are tiny brown salamanders that range in color from reddish to yellowish brown. Some individuals have a stripe down the back to the end of the tail, and a small proportion of the population has no stripe at all. The sides of the stripe are wavy and are usually brighter on the front part of the tail than on the body. The sides have lighter-colored flecks, and the belly is mottled with orange, red, black, and white.

Larvae and Juveniles Recent hatchlings are about 0.5 inches long and resemble adults in coloration.

Other species with similar characteristics and how to distinguish them

Similar Species Webster's salamanders can be confused with zigzag and southern red-backed salamanders, but the ranges of the three species overlap

top Webster's salamanders inhabit rocky areas in hardwood forests with heavy ground litter.

Aquatic Cave Salamanders Four species of eastern salamanders are completely adapted to life in the total darkness of caves.

Stream and Seep Salamanders Of the species of salamanders that are at least partially terrestrial, 36 in the eastern United States breed and stay in the vicinity of small streams or seepage habitats in wooded areas.

Seasonal Wetland Salamanders Among the primarily terrestrial eastern salamanders are 13 species that live in forests but move overland during their annual breeding season to isolated seasonal wetlands such as Carolina bays and vernal ponds.

Forest Terrestrial Salamanders These fully terrestrial species do not require water in which to breed and have no larval stage. The 42 species constitute the highest diversity of this group in the world.

WEBSTER'S SALAMANDER *Plethodon websteri*

Large map showing where the species is found in the eastern United States

Small map showing where the species is found in the continental United States

in only a few areas. Because these species are variable in color pattern, all three must be distinguished by genetic analysis.

Distribution Webster's salamanders are found from South Carolina to Louisiana, but except for a relatively broad area from western Georgia through east-central Alabama they occur as isolated populations.

Habitat Hardwood forests in rocky terrain with heavy ground litter of leaves, logs, and rocks are the preferred habitat.

Behavior and Activity These salamanders are locally abundant on the surface beneath ground litter from October to May but virtually disappear deep underground in summer through early fall.

Subsections on distribution, habitat, and behavior and activity

Webster's Salamander 417

Subsections on reproduction, food and feeding, predators and defense, conservation, and taxonomy of the species

....

Interesting or anomalous facts about the species or about salamanders in general

....

Reproduction Terrestrial courtship and mating generally occur from January to late March, but Webster's salamanders can be found mating as early as late November in southern Mississippi. Females begin laying their clutch of 3–8 eggs underground in June and July. Hatchlings are found above ground in October.

Food and Feeding Webster's salamanders eat small terrestrial invertebrates such as ants, mites, termites, springtails, spiders, snails, centipedes, flies, and worms.

Predators and Defense Specific predators have not been documented but presumably include ringneck snakes, garter snakes, large spiders, and shrews that occupy similar habitat. Observations of populations in southwestern Mississippi provide reliable evidence that predation by turkeys has a significant impact on Webster's salamanders, especially when they are migrating to breeding outcrops. Often the salamanders have missing tails, perhaps due to turkeys targeting the brighter tail than the body.

Conservation No generalized conservation threats have been identified for Webster's salamander other than loss of forested habitat. South Carolina lists the species as Endangered, and Louisiana considers it a Species of Special Concern.

Taxonomy Populations of Webster's salamanders were reported in the scientific literature as early as the 1950s, but as zigzag salamanders. The species was described in 1979, and no subspecies have been recognized.

DID YOU KNOW?
New species of salamanders continue to be described in the eastern United States as DNA analyses differentiate species that look identical. Several new species are expected to be described within the next several years, even as the actual number of salamander populations continues to decline.

aquatic salamanders

AMPHIUMAS

One-toed Amphiuma *Amphiuma pholeter*

Two-toed Amphiuma *Amphiuma means*

Three-toed Amphiuma *Amphiuma tridactylum*

BODY SHAPE	NUMBER OF COSTAL GROOVES	BODY COLOR	BODY SIZE
Long and eel-like; four small legs	One-toed, 65 Two-toed, 57–60 Three-toed, 62	Dark brown or gray to nearly black **PATTERN** Uniform	Adults up to 13 (one-toed), 45 (two-toed), and 41 (three-toed) inches in total length

Description Adults are larger than most other salamanders, and are robust and cylindrical with tiny eyes and four very small limbs. External gills are lacking, but a single pair of gill slits is present. One-toed amphiumas are fairly uniform in body coloration and have a single toe on each of the four tiny legs. Two-and three-toed amphiumas are uniformly dark gray or brown to nearly black with a light gray belly. The body and belly colors are more distinct in the three-toed amphiuma, which also has a dark throat patch that is not present

top Individuals of all three species of amphiumas will bite one another during mating rituals and in male-male aggression. This one-toed amphiuma shows bite-mark scars from such encounters.

Hatchling and juvenile amphiumas have external gills that are absorbed soon after hatching.

in the other two species. Two-toed amphiumas have two toes on each leg, and three-toed amphiumas have three toes.

Larvae and Juveniles Larval amphiumas resemble adults in body form and are black with a light gray belly; they have white gills that are lost shortly after hatching. Juveniles have the same color and pattern as adults.

Similar Species Greater and lesser sirens are long and cylindrical but have large external gills and only one pair of small limbs, which are near the head.

Distribution One-toed amphiumas are found in a limited area in southern Mississippi, extreme southern Alabama, extreme southwestern Georgia, and from the Florida panhandle southward in the Gulf Coast lowlands to Hernando County. Two-toed amphiumas occur in the Coastal Plain from Virginia south through peninsular Florida and west to eastern Louisiana. Three-toed amphiumas occur in the lower Mississippi River drainage in southeastern Missouri, western Tennessee, and eastern Arkansas and from eastern Texas eastward through most of Louisiana, Mississippi, and western Alabama.

Habitat Adult and juvenile one-toed amphiumas occupy swampy and periodically inundated floodplains associated with streams having low to moderate flow. They are also found in mixed bottomland hardwoods and cypress habitats with deep muck derived from hardwood and cypress litter, and in

top Adult amphuima have a single pair of gill slits, tiny eyes, and four very small limbs.

bottom The one-toed amphiuma's gill opening and single toe are apparent on this specimen. The lateral line system, visible in slits on the snout and in pores on the body, senses vibrations in the water.

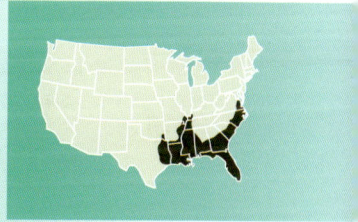

- one-toed amphiuma
  Amphiuma pholeter

- three-toed amphiuma
  Amphiuma tridactylum

- both one-toed and
  two-toed amphiumas

- two-toed amphiuma
  Amphiuma means

pitcher plant bogs. Two-and three-toed amphiumas are generally found in hardwood swamps, cypress swamps, temporary ponds, permanent ponds and lakes, sloughs, canals, and sluggish streams; they sometimes occur in clear, sand-bottomed creeks with little or no detritus on the bottom. They may seek shelter in burrows in the bank.

Behavior and Activity All life history stages are fully aquatic. Amphiumas seldom leave the water but will move short distances over land during heavy rains. They can survive droughts and dry ponds by aestivating in burrows in the substrate. Activity decreases during cold weather.

Reproduction Courtship and mating are completely unknown in the one-toed amphiuma and have not been thoroughly described for the two-and three-toed amphiumas. Females lay strings of 20–200 eggs beneath objects such as logs and other debris in winter and spring, and usually remain coiled around them during incubation. Eggs have been found beyond the water's edge, but these may have initially been laid in water before the water level dropped. Eggs hatch in about 5–6 months; the larval stage lasts about 3 weeks.

Food and Feeding Hatchlings and juveniles probably feed on small invertebrates, but little is known about how they

above Females lay fewer than 200 eggs annually and, like this three-toed amphiuma, live for many years.

right Like the other two species, two-toed amphiumas are fully aquatic predators of a wide array of invertebrates and small vertebrates. They can inflict a nasty wound if they bite you.

Two-toed amphiumas are the longest salamanders in North America. The record total length is more than 45 inches.

capture and handle prey. Adults of the two larger species are powerful predators. Collectively, the three species of amphiumas have been reported to eat crayfish, worms, many insects and their larvae, small frogs and tadpoles, other salamanders, watersnakes, turtles, lizards, small fish, snails, and spiders.

Predators and Defense Amphiumas are the primary diet of adult mud snakes, but cottonmouths, kingsnakes, some watersnakes, alligators, wading birds, sandhill cranes, and otters will eat them as well. Feral hogs rooting in the mud of drying ponds and lakes dig amphiumas out of the bottom and may constitute a major source of mortality. The larvae and small juveniles are potential prey of predatory insects, small aquatic snakes, and fish. Large amphiumas bite hard and can inflict a serious wound on potential predators, including humans. They also produce slimy mucus that makes them very slippery.

Conservation The one-toed amphiuma is rare throughout its range and is of conservation concern in all states where it occurs; Alabama considers it a Species of High Conservation Concern. Two-and three-toed amphiumas are widely distributed, often in large populations, and are not of serious conserva-

above The one-toed amphiuma is rare throughout its range and is of conservation concern in all states where it occurs

left The contrast between upper and lower body colors is more distinctive in the three-toed amphiuma.

tion concern in most areas. Feral hogs, an introduced species, may consume significant numbers of amphiumas when ponds and lakes are drained, and the effects of their predation deserve further examination.

Taxonomy Three species of amphiumas are recognized in the eastern United States. The one-toed amphiuma was not described until 1964. Two-toed amphiumas were described in 1821; three-toed amphiumas were described in 1827. No subspecies are recognized. Congo eel and lamp eel are common names in some regions.

HELLBENDER *Cryptobranchus alleganiensis*

BODY SHAPE
Large; flattened top to bottom; head shape broad and flattened with small eyes

NUMBER OF COSTAL GROOVES
Costal grooves inconspicuous and hidden within skin wrinkles

BODY COLOR
Light brown, orangish, or nearly black

PATTERN
Faded dark spots or blotches on a lighter background

BODY SIZE
Adults up to 29 inches maximum total length, but most adults a little over 12 inches long

Description The fully aquatic adult is large and flattened with wrinkled folds of skin on each side of the body. The head is very broad and also flattened, and the eyes are small. The paddle-like tail is broad and flat. Hellbenders can range from brownish to bright orange; some have small spots or irregularly shaped blotches. The belly is usually plain.

Larvae and Juveniles Larvae are chunky, have bushy gills, and develop irregular spots as they mature. They absorb their external gills after about 2 years when they reach 4–5 inches in length and reach maturity within 5–8 years. Juveniles resemble adults in color and pattern.

top This hellbender in a mountain stream has the species' characteristic broad, flattened head and body, paddle-shaped tail, and wrinkled folds of skin on the sides.

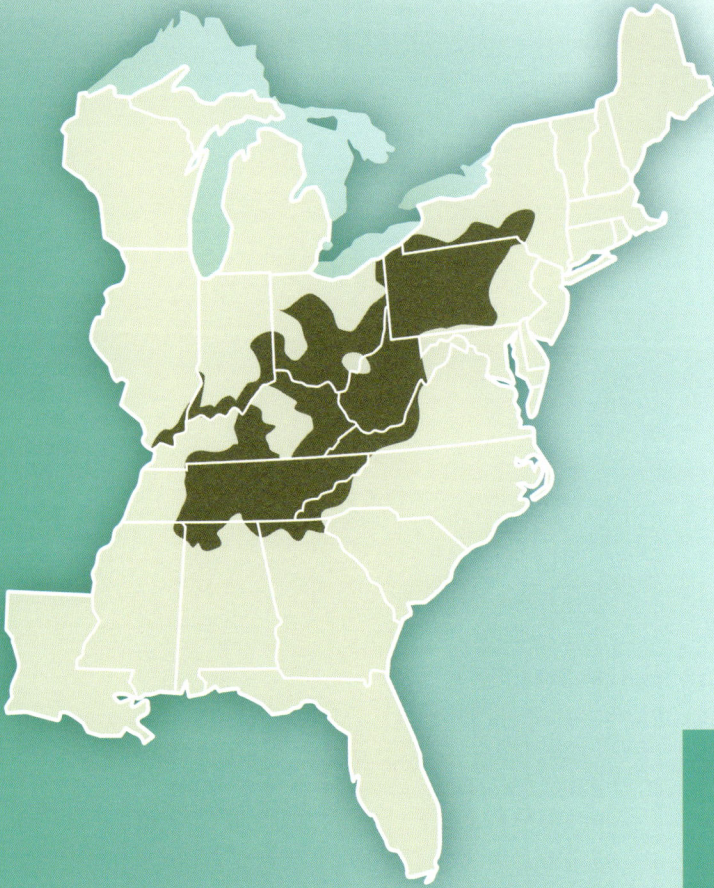

Similar Species The broad, flat body of hellbenders readily differentiates them from mudpuppies, the only other large salamanders whose geographic range overlaps theirs. Mudpuppies also have obvious fluffy gills as adults.

Distribution Hellbenders occur in extreme northeastern Mississippi, extreme northern Alabama, and Georgia (within the Tennessee River drainage); much of Kentucky and Tennessee; western North Carolina; and southwestern Virginia. They range up the Appalachians through Pennsylvania into New York. They are also found in the Ohio River drainage of Ohio, Indiana, and extreme southern Illinois. Except for a few streams in the Northeast, they occur only

left Larvae have bushy gills and develop irregular spots as they mature.

below Hellbenders' sharp, serrated teeth can result in a nasty bite.

in drainages that eventually empty into the Mississippi River. They also occur in southeastern Missouri and in several disjunct localities in northeastern Arkansas.

Habitat Hellbenders occupy rivers and large streams, and also small mountain streams as long as these are permanent. They prefer riffle zones with large, flat rocks but may be in any portion of the stream that offers cover.

Behavior and Activity These impressive salamanders rarely leave the water. They emerge from their rocky hiding places at night to search for food, but they can be very active in the daytime during the breeding season (September and Oc-

DID YOU KNOW?

Most salamanders cannot open their mouth wide enough to bite humans, but amphiumas and hellbenders can give a nasty bite with their sharp, serrated teeth.

tober). In some streams they are diurnal during much of the summer. Males are territorial during the breeding season and defend nesting rocks from other hellbenders, but even outside the breeding season a single rock seldom shelters more than one. Hellbenders generally move only short distances and maintain small territories of a few hundred square feet, but they have been known to move several miles within a river. They are very slimy and therefore very difficult to catch and hold. The natural life span may be 30 years or more.

Reproduction Females lay 200–400 or more eggs, usually between August and October, in depressions the male makes under his rock. A male may mate with and fertilize the eggs of several females. The male guards the eggs until they hatch some 2 months later. Some males will allow their young to remain under their rock for 6 months or more. Nesting habitat (large rocks not covered in silt) seems to be a limiting resource in most streams.

above Hellbenders may be caught by fishermen. They are often killed as a result or let go with the hooks still in their mouths.

left Without the normal color and pattern to provide camouflage, albino hellbenders are more easily seen and captured by predators.

Food and Feeding Hellbenders eat crayfish primarily, and also small fish (including lampreys), aquatic insects, salamanders, and occasionally other hellbenders. Game fish are too fast for hellbenders to catch and are not typical food items. During the breeding season, their own eggs may comprise a major component of their diet.

Predators and Defense The common enemies of adults include humans, who hook hellbenders on lines while fishing or trapping and kill them. Large fish, some turtles, and northern watersnakes eat hellbenders; otters probably are natural predators as well. Larvae and small juveniles are presumably eaten by large predatory insects and fish as well as other hellbenders. These rather intimidating salamanders sometimes bite if handled but are not venomous. Their main defenses against predators

opposite Hellbenders generally move only short distances, and males ferociously defend their territory especially when guarding eggs. This photo shows the color variation within the same stream in northern Georgia

are concealment under rocks, their camouflage, and the noxious mucus that covers their body and makes them difficult to catch and handle.

Conservation Hellbenders no longer occupy much of their historic range because siltation from land erosion has destroyed their habitat. Damming of rivers has also eliminated much of the hellbender's original habitat, especially in the Tennessee River valley. Forest cover along streams and rivers is important for maintaining healthy water for hellbenders. The hellbender is a Species of Special Concern in North Carolina and Virginia, is listed as Threatened by the state of Georgia, and is deemed in need of management in Tennessee. It is very rare in Alabama, where it is classified as a Species of Highest Conservation Concern. The Ozark subspecies is federally listed as endangered. Anglers can affect local hellbender population levels in any stream by deciding whether to kill these salamanders or release them. Habitat management programs to conserve and protect the rivers and streams in which this unique, long-lived salamander occurs are needed for conservation efforts. Accurate data on population sizes

throughout the geographic range would allow a better evaluation of the species' conservation status.

Taxonomy The eastern hellbender (*Cryptobranchus alleganiensis alleganiensis*) was first described in 1803. Other common names are grampus, alligator, snot otter, devil dog, mud devil, mollyhugger, and giant salamander. Hellbenders are also often erroneously called mudpuppies or waterdogs. The Ozark population (*Cryptobranchus alleganiensis bishopi*) has at times been elevated to the species level.

Hellbenders generally move only short distances and maintain small territories of a few hundred square feet, but they have been known to move several miles within a river.

SHOVEL-NOSED SALAMANDERS

Shovel-nosed Salamander   *Desmognathus marmoratus*

Golden Shovel-nosed Salamander   *Desmognathus aureatus*

Western Shovel-nosed Salamander   *Desmognathus intermedius*

BODY SHAPE
Large; head flattened top to bottom; tail flattened side to side

NUMBER OF COSTAL GROOVES
14

BODY COLOR
Dark brown, gray, or black

PATTERN
Two rows of yellowish brown, gray, olive, or yellowish spots

BODY SIZE
Adults up to 5.75 inches in total length

Description   Shovel-nosed salamanders are large and robust. Their common name comes from the slanted, wedge-shaped snout. The slant starts well behind the small eyes on the flattened head, while the slope in other salamanders found in the same region starts in front of the eyes. Body coloration consists of two rows of highly irregular yellowish brown, gray, olive, or yellowish spots over a dull, dark brown, gray, or black background that provides excellent camouflage

top Shovel-nosed salamanders remain in water in southern Appalachian mountain streams for most of their lives.

in the rocky streams in which this salamander lives. Two rows of tiny light spots occur on each side of the body in some populations. The belly is gray and may either lack a pattern or have dark mottling; it darkens in the center with age. The tail is flattened side to side and has a sharp upper keel.

Larvae and Juveniles Hatchlings have two rows of light spots on a gray back. They are nearly black and have whitish gills. Other *Desmognathus* larvae that co-occur with them are not as dark. Older juveniles are lighter and have noticeable light flecks on each side, and their legs are longer than those of adults.

Similar Species The black-bellied salamander may be confused with the shovel-nosed, but the former's uniformly black belly, bulging eyes, and more conspicuous eye-to-jaw stripe are distinctive. All other members of the genus *Desmognathus* have two round openings inside the roof of the mouth (internal nares); shovel-nosed salamanders have only slits. Although they may be difficult

Juvenile shovel-nosed salamanders bear pairs of light spots that disperse as they attain maturity.

● shovel-nosed salamander
  Desmognathus marmoratus

● western shovel-nosed
  salamander
  Desmognathus intermedius

● golden shovel-nosed
  salamander
  Desmognathus aureatus

to see, the internal nares may be visible in live salamanders if the mouth can be pried open carefully, but this is not recommended.

Distribution   Species in the shovel-nosed salamander complex occur in the southern Appalachian Mountains from southwestern Virginia through western North Carolina and extreme eastern Tennessee southward into northeastern Georgia and extreme northwestern South Carolina. The golden shovel-nosed salamander (*Desmognathus aureatus*) is found in northeast Georgia and nearby North and South Carolina. The western shovel-nosed salamander (*Desmognathus intermedius*) is found on the borders of North and South Carolina in and around

Shovel-nosed salamanders inhabit cool, rocky streams where their dull brown and yellowish body pattern serves as excellent camouflage.

the Smoky Mountains. The shovel-nosed salamander (*Desmognathus marmoratus*) ranges north of the two newly described species and occurs in northwestern North Carolina, western Virginia, and West Virginia.

Habitat Cool, highly oxygenated streams above 300 m elevation that have loose gravel, rocks, and moderate flow gradients are preferred. Adults are more likely to inhabit riffles than slower pools because disturbed water contains more oxygen and improves respiration across the skin. Larvae remain in the interstices of rocks, gravel, and leaves on the bottom of streams well away from adults.

Behavior and Activity Adult and juvenile shovel-nosed salamanders remain in water for most of their lives, although some adults emerge from their streams on wet nights and move a short distance into the surrounding forest where they remain under litter during the day. They can be found in running water every month of the year. Shovel-nosed salamanders are not known to migrate, and little is known about their movement activities.

Reproduction The mating behavior has not been described, but presumably mating occurs in underwater riffles during late winter and early spring. Females lay clutches of 20–65 eggs in late spring and summer. Eggs are attached singly or in clusters of 2–4 on the underside of large rocks in flowing water. Hatching occurs from mid-August through September. Larvae may remain in the larval stage for as long as 3 years before absorbing their gills and going through metamorphosis.

Food and Feeding Insects that live in the water or fall into it are the primary prey of adults and juveniles, which also eat snails, crayfish, beetles, flies, mayflies, caddisflies, wasps and bees, and other salamanders such as two-lined salamanders that live in the same streams. Adults occasionally eat juveniles and larvae of their own species. The larvae eat a wide variety of aquatic insects. Feeding takes place at night, usually while the salamanders are hiding beneath stones and watching for prey to swim by or while they are perched on rocks.

Predators and Defense Predators of shovel-nosed salamanders include several species of carnivorous fish such as sculpin and brook trout, northern watersnakes, and large stonefly nymphs. Some individuals practice cannibalism. Fleeing for the shelter of stones and leaves on the stream bottom is the primary means of escaping predation.

Conservation The high-elevation streams used by these salamanders are susceptible to siltation from upslope sources such as road building, forest clearing, and home construction. Their need for highly oxygenated streams makes them sensitive to pollution of all sorts and to siltation from upstream

Body coloration consists of two rows of highly irregular yellowish brown, gray, olive, or yellowish spots over a dull, dark brown, gray, or black background.

activities and dam construction. Like other stream salamanders, shovel-nosed salamanders are collected and used for fish bait, but less commonly now than a few decades ago.

Taxonomy The shovel-nosed salamander was described in 1899 and remained in its own genus (*Leurognathus*) for many years until genetic evidence indicated that this species is closely related to the *Desmognathus* salamander group. Its genetic relationship with the black-bellied salamander is unclear. High levels of variation in color and pattern across the range and in individual populations suggest that there may be as many as 3–4 cryptic species within the shovel-nosed salamander complex, and in 2023 this complex was delineated into three species instead of one based on both mitochondrial and nucleic DNA differences: the golden shovel-nosed salamander (*Desmognathus aureatus*); the western shovel-nosed salamander (*Desmognathus intermedius*); and the shovel-nosed salamander (*Desmognathus marmoratus*).

MANY-LINED SALAMANDER *Stereochilus marginatus*

BODY SHAPE
Slender with a
short tail

**NUMBER
OF COSTAL
GROOVES**
18

PATTERN
Broad line along
back and numerous
thin lines on sides

BODY SIZE
Adults up to 4.5
inches in total
length

BODY COLOR
Brown and yellow

Description The adult is a small, slender salamander with a relatively short tail. A dark brown band on the back runs through the eye and onto the tail. The sides have a series of fine, black longitudinal lines or streaks that may appear as a series of spots. The belly is yellow with black flecks. The tail is flattened side to side toward the tip and has a keel.

Larvae and Juveniles The larvae and juveniles are distinctly dark brown above and yellow below. The head and body may have small yellow or white spots. Older larvae may be mottled on the back.

Similar Species Dwarf salamanders are smaller and have a broad brown stripe down the back bordered by black lines, a long tail, no thin lines on the sides,

top Adult many-lined salamanders are distinctive in having a dark brown band that runs through the eye onto the back and black streaking on the body.

top Large many-lined salamander larvae have dark backs and yellow bellies.

bottom Small larvae may be reddish and have distinctive spots on the body and tail.

Many-lined salamanders inhabit swamps, ditches, and other wetlands containing sphagnum and other dense vegetation.

and only four toes on the rear feet. Mud salamander larvae are light brown with a few black spots rather than dark brown with yellow spots.

Distribution Many-lined salamanders are found in the Coastal Plain from southeastern Virginia to northern Florida.

Habitat These salamanders occupy gum and cypress swamps, woodland ponds, ditches, canals, sluggish streams, and other shallow permanent wetlands. The mucky pools, sphagnum mats, and leaf litter often found in these wetlands are used extensively for concealment. All life stages occupy shallow, acidic waters.

Behavior and Activity Little is known about the seasonal activity patterns and movements of this secretive salamander. Individuals spend their entire lives in small patches of wetlands, aestivating in small cavities in the substrate when the water dries up and sometimes hiding under logs and other surface objects adjacent to wetlands.

Reproduction Reproduction takes place in water in the fall. Courtship and mating behaviors are incompletely described but are probably similar to those of other members of the plethodontid group; males produce spermatophores that are picked up by females for internal fertilization. Females lay 16–121 eggs in water or on nearby land in the winter, attaching them singly to the underside of logs, debris, or other objects, or depositing them in aquatic mosses in

Loss of wetlands is a major threat to the many-lined salamander.

shallow water. Females do not attend their eggs. Incubation lasts 1–2 months, and eggs hatch in late March and April.

Food and Feeding The available information indicates that many-lined salamanders eat a diverse selection of prey, including small clams, amphipods, isopods, and insects and their larvae.

Predators and Defense Natural predators have not been documented but likely include predatory aquatic insects, snakes, fish, and wading birds. Defense mechanisms or behaviors are unknown.

Conservation The primary threat to this salamander is loss of wetlands to drainage and conversion to urban, agricultural, and industrial uses. It is a Species of Special Concern in Virginia because its distribution is limited and the wetland habitats it requires are being destroyed.

Taxonomy The many-lined salamander was first described in 1856. No subspecies are recognized.

BLACK WARRIOR WATERDOG
or ALABAMA WATERDOG *Necturus alabamensis*

BODY SHAPE
Robust; big, feathery gills; laterally compressed tail

NUMBER OF COSTAL GROOVES
16

BODY COLOR
Gray to brown to almost black

PATTERN
Irregular black spots

BODY SIZE
Adults up to 8.5 inches in total length

Description This species, like all *Necturus*, has feathery, external gills in adults (neotenic); pigmented skin; four well-developed limbs with four toes on both the front and hind feet; and a laterally compressed tail. Adults can reach 6–8.5 inches in total length. The dorsum is reddish brown to almost black with some populations having small dorsal spots. The belly lacks spots, and the tips of the toes are light colored. The body and head are flattened. A dark eye stripe runs from the nostril through the eye to the gills.

Larvae and Juveniles Larvae are 1.1–2 inches in total length and are striped with large dorsal spots on a reddish-brown dorsum. Juveniles can have light stripes on the head and back and, as with the adults, generally have a dark eye stripe.

top The Black Warrior waterdog is found in a restricted area of northern Alabama and is afforded federal protection.

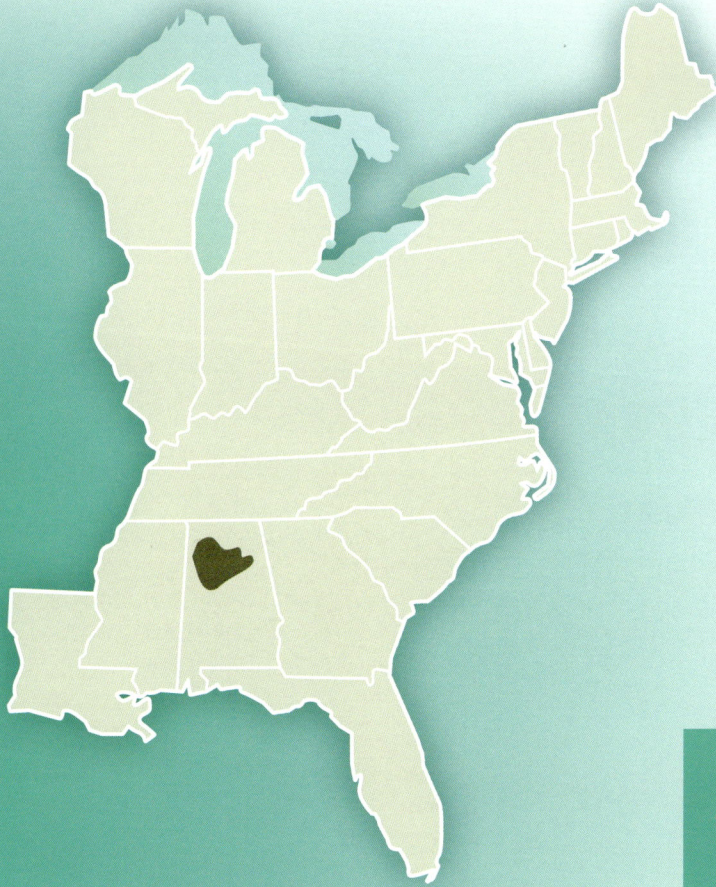

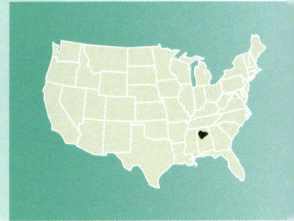

Similar Species The only other *Necturus* species that can be found in the same habitat is N. *beyeri*, which is more cylindrical, has belly spots, and lives in burrows rather than under rocks or debris. Other sympatric salamanders have five toes on their back feet.

Distribution Black Warrior waterdogs are found in north and north-central Alabama in the streams of the upper Black Warrior River drainage, including parts of the North River, Locust Fork, Mulberry Fork, and Sipsey Fork.

Black Warrior River waterdog larva

Habitat Black Warrior waterdogs seem to prefer streams with logjams, woody and/or leaf debris, or rocks that provide cover and locations for egg laying. Their flattened body is likely correlated to their habit of sheltering under rocks or other cover in stream beds.

Behavior and Activity Relatively little is known of the natural history of Black River waterdogs. They are seldom seen during the summer, and most of their activity is centered around their breeding season from December to February. They are primarily nocturnal in their activity.

Reproduction Adults mate from late fall through the winter. Females attach 4–40 eggs to the underside of logs, rocks, and other objects on the stream bottom in April and May. Males often guard the nest until the eggs hatch in June, July, or August.

Food and Feeding Adults are opportunistic, eating most small animals they can capture and swallow. Their diet includes many types of insects (including their larvae and nymphs), clams, snails, earthworms, slugs, leeches, spiders, crayfish, and small slow-swimming fish.

Predators and Defense Watersnakes are known predators, and crayfish and fish are likely predators as well. Black Warrior waterdogs are inactive during the summer months, which may help keep them hidden from predation. Skin secretions may deter some predators.

Conservation The highly permeable skin and external gills of the waterdog make it particularly sensitive to declines in water quality and oxygen concen-

Black Warrior River waterdogs are brownish with a few black spots on the back and a uniformly gray belly. All waterdogs have feathery gills and a large, side-to-side-flattened tail.

tration. In 2018 the U.S. Fish and Wildlife Service (USFWS) listed the Black Warrior waterdog as Endangered. Low and declining population numbers due to habitat loss and fragmentation and poor water quality led to the designation of 420 river miles of critical habitat in the Black Warrior River Basin, comprising more than 50 percent of the waterdog's historical habitat and including 127 miles of habitat already designated for other federally protected fish, mussels, and salamanders.

Taxonomy The taxonomy of the Black Warrior waterdog has historically been problematic. It coexists in a few localities with the Gulf Coast waterdog but is distinct in microhabitat preferences, morphology, and genetics. It was originally described in 1937 but better defined in 2003.

BODY SHAPE
Robust; feathery external gills; flattened tail

NUMBER OF COSTAL GROOVES
14

BODY COLOR
Rusty brown

PATTERN
Numerous irregular brown to black spots

BODY SIZE
Adults up to 9 inches in total length

Description The Neuse River waterdog is a large, stout salamander with feathery gills, four toes on all four feet, and a large tail that is laterally flattened. It has a rusty brown body with numerous large dark spots or blotches across the body and tail. The adult has a brown to gray belly with smaller spots.

Larvae and Juveniles Larvae are stout and tan with a dorsal stripe, and the sides are darker with pale flecks. The back and belly are spotted on the juveniles.

Similar Species Dwarf waterdogs are sympatric with Neuse River waterdogs but usually lack markings on their dark gray body and have a white, unspotted line down the middle of their belly. All of the mole salamander (Ambystoma) larvae also have feathery gills but are small, do not occur in rivers, and have five toes on the hind feet.

Distribution Neuse River waterdogs inhabit the Neuse and Tar River systems of North Carolina.

top Neuse River waterdogs have large bluish black blotches on the brown body and tail.

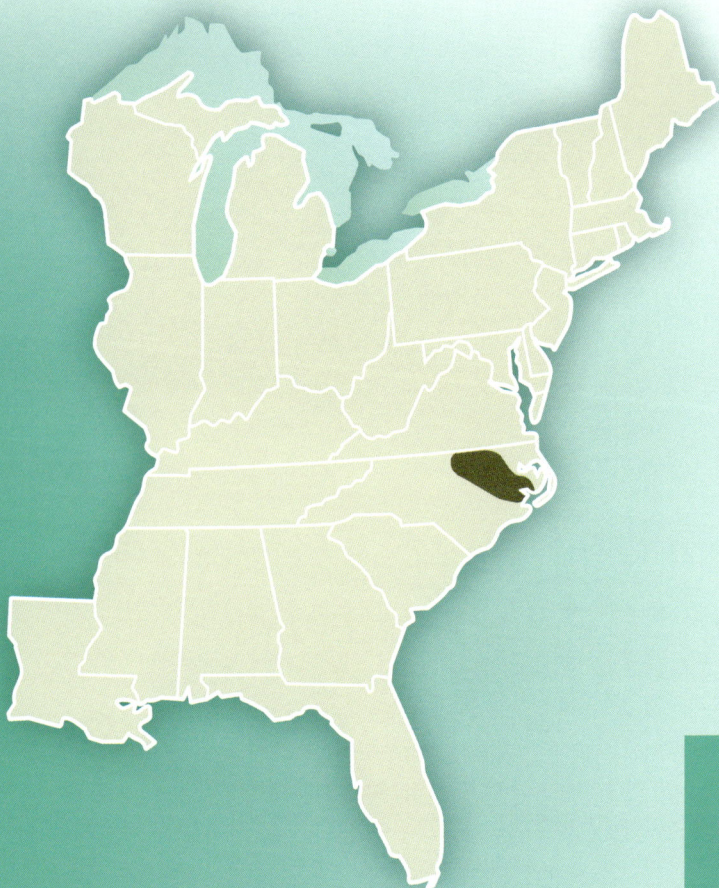

Habitat Both larvae and adults of this permanently aquatic species occur in detritus, especially leaf beds, and logjams within rivers and streams. The Neuse River waterdog does occur in streams with clay or hard soil substrate.

Behavior and Activity Neuse River waterdogs are more active in the cooler months of fall and spring than in winter or summer. Activity increases after moderate rainfall as food becomes dislodged but ceases during heavy rains.

Reproduction Females deposit eggs during the summer in streams and small rivers. They lay up to 35 unpigmented eggs measuring 0.31–0.35 inches under-

Neuse River waterdogs occur only in the Neuse River watershed in North Carolina.

neath submerged objects. Larvae hatch at 0.83–0.94 inches in total length. The young reach maturity at six years of age.

Food and Feeding Adults are opportunistic, eating most any small animals they can capture and swallow. Their diet includes many types of insects (including their larvae and nymphs), clams, snails, earthworms, slugs, leeches, spiders, crayfish, and small slow-swimming fish.

Predators and Defense Watersnakes are known predators, and crayfish and fish are likely predators as well. Neuse River waterdogs are inactive during summer months, which may help keep them hidden from predation. They have skin secretions that may deter some predators.

Conservation The Neuse River waterdog is listed as a Species of Special Concern in North Carolina due to its restricted range and vulnerability to siltation, pollution, and stream alterations.

Taxonomy The Neuse River waterdog was first described in 1924.

GULF COAST WATERDOGS

Gulf Coast Waterdog   *Necturus beyeri*

Apalachicola River Waterdog   *Necturus moleri*

Escambia River Waterdog   *Necturus mounti*

BODY SHAPE
Robust; big feathery gills; flattened tail

NUMBER OF COSTAL GROOVES
16–17, rarely has 18

BODY COLOR
Gray to brown to black

PATTERN
Irregular spots

BODY SIZE
Adults up to 11 inches in total length

Description Adult Gulf Coast waterdogs are large, robust salamanders with four toes on each of their four feet; large, feathery red gills; and a large, flattened tail. Three species occur along the Gulf Coast of the Southeast. The Gulf Coast waterdog (N. beyeri) is a large (9.5–11 inches in total length), brown, aquatic salamander with numerous white spots across the entire body. Dark brown spots, which appear overtop the pattern of small white spots, increase in size and darkness from head to tail. The head lacks a dark stripe from the nostril through the eye (although dark brown spots may partially occupy this area).

top An adult Gulf Coast waterdog

Gulf Coast waterdog larvae

The center of the brown-to-tan belly lacks the small white spots but has small brown spots that become larger along the sides. The Escambia River waterdog (N. *mounti*), is smaller in size (6.4–6.5 inches in total length), possesses dark dorsal and side spotting with the spots only about the size of its eye, and lacks spotting on the chin or belly. The Apalachicola River waterdog (N. *moleri*) is also small in size compared to N. *beyeri* (6.5–7 inches in total length), possesses small dark dorsal and side spotting, and has some spotting on the chin and lateral portion of the belly.

Larvae and Juveniles Larvae of all three species have numerous small spots and lack dark stripes. Juvenile N. *beyeri*, like the larvae, have small white spots, while both N. *mounti* and N. *moleri* lack these white spots.

Similar Species All mudpuppies and waterdogs are large aquatic salamanders with feathery red gills, a flattened tail, and four

left All waterdogs have four toes on each foot, as seen in this Escambia River waterdog from Okaloosa County, Florida.

below Escambia River waterdogs become very inactive during the hot summer months.

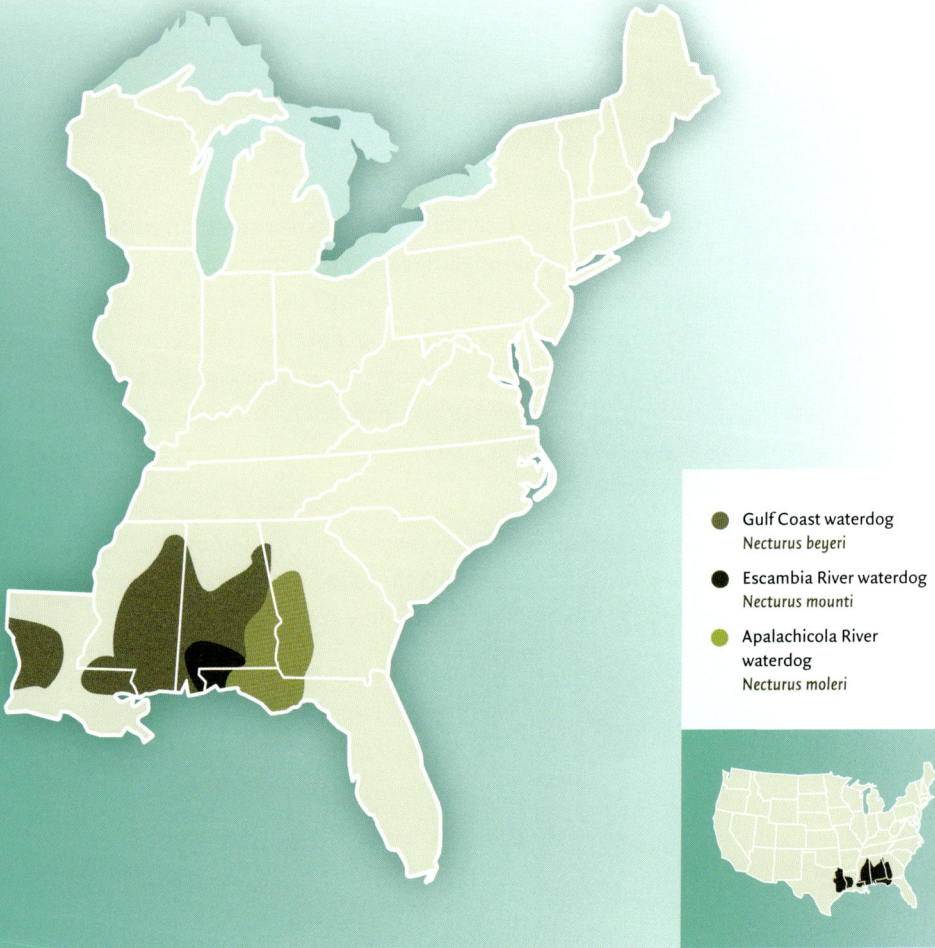

Gulf Coast waterdog
Necturus beyeri

Escambia River waterdog
Necturus mounti

Apalachicola River
waterdog
Necturus moleri

toes on all feet, whereas all mole salamanders are smaller, rarely in streams, and have five toes on the back feet. The dwarf waterdog usually lacks markings on a gray body and has a white line down the middle of its belly. Mudpuppies have a conspicuous dark line through each eye and irregular black spots on a gray-to-tan body and tail. Both the Escambia River and Apalachicola River waterdog larvae and adults lack white spotting and are confined to their unique drainages.

Distribution The Gulf Coast waterdog (N. beyeri) is found in the Mobile River drainage and north to the Alabama-Georgia border on the Coosa River. They also

above and left The Apalachicola waterdog is a recently described species. Most waterdogs have large, feathery gills, an adaptation to slow-moving Coastal Plain streams with low levels of dissolved oxygen.

occur in northeastern Mississippi and in southern Mississippi to southeastern Louisiana, skipping the Mississippi River drainage but with a disjunct enclave occuring in southwestern Louisiana and southeastern Texas. The Escambia River waterdog is confined to the Escambia River drainage, including the Blackwater, Escambia (Conecuh in Alabama), Perdido, and Yellow River drainages of Alabama and the panhandle of Florida. The Apalachicola River waterdog is restricted to the Apalachicola, Chipola, Choctawhatchee/Pea, Econfina, and Ochlockonee River drainages of Alabama, Florida, and Georgia.

Habitat Both juveniles and adults prefer logjams, detritus, and leaf litter in the streams and rivers that they inhabit. Waterdogs of all ages use natural depressions and crayfish burrows under stream banks for shelter.

Behavior and Activity Waterdogs are inactive during hot weather, with activity peaking during the fall and winter. They are active after moderate rains but cease activity during heavy rain. They forage primarily at night and remain hidden in leaf litter, leaf beds, and stream-bottom debris during the day. Adults in burrows have been reported to carry gravel and sand to the burrow openings. Both males and females defend their burrows.

Reproduction Gulf Coast waterdogs mate in water from late fall through winter. Females attach 4–40 eggs in oblong clusters to the undersides of logs, rocks, and other objects in April and May.

Food and Feeding Adults and juveniles are opportunistic, eating most any available prey items that they can swallow. Prey items include many types of insects, clams, snails, worms, leeches, spiders, crayfish, and small fish.

Predators and Defense Watersnakes, fish, crayfish, and turtles, especially the alligator snapping turtle, are common predators of these species. Burrowing and summer inactivity may be ways that Gulf Coast waterdogs hide from potential predators.

Conservation Gulf Coast waterdog populations appear to be stable but any aquatic salamander is always susceptible to alteration of streams, pollution, and siltation.

Taxonomy The taxonomy of Gulf Coast waterdogs has been debated since N. beyeri was first described in 1937. The currently accepted taxonomy has redefined N. beyeri and described two new species: N. mounti and N. moleri. Changes and/or additions to these new species are possible, especially with regard to both the western enclave and the Mobile River drainage lineages of N. beyeri.

DWARF WATERDOG   *Necturus punctatus*

BODY SHAPE	**NUMBER OF COSTAL GROOVES**	**PATTERN**	**BODY SIZE**
Slender; bushy gills; flattened tail	14–16	Uniformly colored, without spots	Metamorphs 2 inches; adults up to 7.5 inches in total length
	BODY COLOR		
	Brown to gray		

Description Dwarf waterdogs are the smallest and most slender of the water-dog group. The body is usually a uniform slate gray to brown without spots, although populations in the Cape Fear and Lumber River drainages in North Carolina have some irregular dark spotting on the tail. Each of the four feet has four toes. The large, red, bushy gills are conspicuous. The midline of the belly is pale cream to white without spots.

Larvae and Juveniles The larvae are brown but otherwise resemble adults. Hatchlings and juveniles are uniformly brown without spots or stripes. The tail fin may be mottled, and the belly is bluish white.

Similar Species All other waterdogs and the mudpuppy are larger, more robust, and have irregular spots as adults. Juvenile dwarf sirens may have

top This dwarf waterdog from Aiken County, South Carolina, has the large gills and drab, un-marked body typical of the species.

The dwarf waterdog has conspicuously large, red, bushy gills.

longitudinal light and dark stripes. Neuse River waterdogs have a rusty brown body, dark spots or blotches on the body and tail, and a dull brown to gray belly with spots.

Distribution Dwarf waterdogs occur in the Coastal Plain and lower Piedmont from southern Virginia through the Carolinas to central Georgia. They are rare to absent from the mainstream Neuse and Tar rivers in North Carolina.

Habitat Sluggish streams with an accumulation of silt, debris, and leaves appear to be preferred, although this may reflect modern impacts to the streams because dwarf waterdogs can also be abundant in clear-flowing blackwater streams. Adults and juveniles are most often found in leaf beds in shallow water.

Behavior and Activity Winter appears to be the primary season of activity, but dwarf waterdogs can be found at all times of the year. Individuals may aggregate in leaf beds in the winter. Adults, larvae, and juveniles are apparently nocturnal, and all life stages are aquatic.

Reproduction Courtship and mating have not been described but probably

Juvenile dwarf waterdogs are uniformly brown but may have dark mottling on their tails.

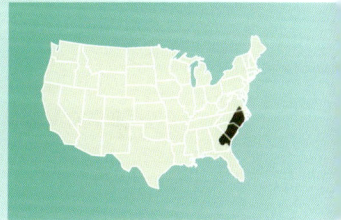

resemble the behaviors observed in other waterdogs. Mating probably occurs in winter, with egg deposition in March to May. Females lay 15–55 eggs, presumably under objects in the stream as other waterdogs do, although no nests have been described. Larvae have been caught in April and November, but the length of the larval period is unknown.

Food and Feeding Adults and juveniles are generalist predators that will eat any animal they can fit into their mouth. Worms, insects and their larvae, and aquatic invertebrates have been identified as prey.

Some dwarf waterdog populations in the Cape Fear and Lumber river drainages in North Carolina have some irregular dark spotting on the tail.

Predators and Defense Predators presumably include some fish, otters, wading birds, sirens, and watersnakes; and may also include other dwarf waterdogs. Dwarf waterdogs are not known to produce noxious skin secretions.

Conservation Although the species is apparently not common anywhere, dwarf waterdogs are not protected in any state in which they occur. Stream pollution and siltation from agricultural, residential, and industrial sources undoubtedly affect this gill-breathing salamander. The conservation status needs modern evaluation throughout the species' range.

Taxonomy The dwarf waterdog was first described in 1850. No subspecies are recognized.

MUDPUPPY *Necturus maculosus*

BODY SHAPE	NUMBER OF COSTAL GROOVES	PATTERN	BODY SIZE
Robust; bushy gills; flattened tail	None	Irregular spots or blotches on body and tail	Adults up to 19 inches in total length
BODY COLOR			
Gray to dark brown			

Description Mudpuppies are large aquatic salamanders with large, dark red, feathery gills; four limbs with four toes on each foot; and a large, flattened, paddle-like tail. Adults are usually rusty brown or occasionally grayish, with irregular bluish black spots or blotches on the body and tail. The belly is also gray with black spots. A dark line passes from the nostril on the flattened snout to the gills on each side of the head. Individuals of the Red River mudpuppy subspecies have a dark stripe down the back bordered on each side by a faint light stripe and large spots or blotches on the body. The belly is light gray with no spots.

Larvae and Juveniles Larvae have a broad, dark line down the middle of the back that is bordered on each side by a yellow stripe. Juveniles have the

top This mudpuppy from Kentucky has the dark body spots characteristic of the species and a dark line from the snout to the gills.

Larvae and juveniles have a broad, dark line down the middle of the back that is bordered on each side by a yellow stripe.

same pattern as larvae, but the yellow and black stripes on the sides are more pronounced.

Similar Species The only waterdog that may occur in the same geographic area is the Black Warrior River waterdog, which is reddish brown to nearly black and usually has no black spots on the back and a uniformly gray belly. All other waterdogs occur well outside the known range of the mudpuppy. The hellbender has a broad, flat body with wrinkled folds of skin.

Distribution Mudpuppies range from southern Quebec to southeastern Manitoba in Canada and south to eastern Kansas and eastern Tennessee in the United States. They also occur across the Midwest from Minnesota to New York, including Wisconsin, Illinois, Indiana, and Ohio, and along the Great Lakes through eastern Pennsylvania, Vermont, Massachusetts, and Connecticut. In

left Mudpuppies have large, dark red, feathery gills, especially in ponds and lakes.

right The dark stripe passing through the eye is pronounced on most individuals. Gill size may be reduced in highly oxygenated water and in captivity.

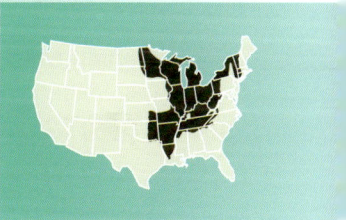

the Southeast, they are found in Kentucky; southwestern Virginia; Tennessee; western North Carolina; northern Alabama, Georgia, and Mississippi; and northern Louisiana. The Red River mudpuppy occurs from north-central Louisiana west of the Mississippi River to southeastern Kansas.

Habitat Adults occupy clear to silted lakes, reservoirs, ditches, streams, and rivers in the Southeast. Flat rocks, logs, debris, crayfish burrows, undercut banks, and rock slabs provide hiding places. Water with high oxygen content below riffles is preferred. Juveniles inhabit pools with a substrate of litter and organic debris.

This is the Red River subspecies of the mudpuppy.

Behavior and Activity Mudpuppies are active in all months of the year and probably remain in the same vicinity for most of their long lives (as long as 34 years). Individuals occasionally move upstream and into tributaries.

Reproduction Mating probably occurs from fall through early spring in most of the Southeast. Males search for females in shallow-water retreats. The male courts a female to entice her to pick up a sperm packet that he has deposited on the substrate. Females turn upside down to attach each of their 20–200 eggs singly to the roof of a nesting chamber excavated beneath a rock, log, or other structure embedded on the stream bottom at a depth of less than 2 feet. Hatching occurs 1.5–2 months later in early summer, but the hatchlings remain in the chamber until the external yolk sac is absorbed, as long as 6–8 weeks. Emergence from the nest chamber usually occurs in August. Females often guard the eggs until they hatch.

Food and Feeding Mudpuppies eat almost any creature they can get into their mouth; only the size of their gape limits what they can consume. Crayfish and small fish are important components of the diet, but a wide variety of insects and their larvae, worms, snails, some amphibians such as dusky and two-lined salamanders, small turtles, fish eggs, and eggs of other mudpuppies are also eaten.

Predators and Defense Animals that eat mudpuppies include crayfish, predatory fish, carnivorous turtles, watersnakes, herons, otters, and larger mudpuppies. The larvae and smaller individuals may also become food for predatory larvae of aquatic insects. Defense is usually a quick escape with a rapid movement of the flattened tail.

Conservation Pollution and siltation from agricultural, residential, and industrial sources threaten streams and their salamander populations throughout the range of this species. The mudpuppy is a Species of Special Concern in North Carolina. The ecology and life history have been well studied in the northern portion of the range, particularly in the Great Lakes area, but little is known about this species in the Southeast.

Taxonomy Two recognized subspecies are found in the Southeast. The common mudpuppy (N. m. maculosus), described in 1818, is found from Kentucky and southwestern Virginia south to northern Georgia, Alabama, and Mississippi. The Red River mudpuppy (N. m. louisianensis), described in 1938, is in northern Louisiana. Some researchers elevate this subspecies to the species level, but currently the accepted taxonomy keeps it as a subspecies.

DWARF SIRENS

Northern Dwarf Siren *Pseudobranchus striatus*

Southern Dwarf Siren *Pseudobranchus axanthus*

BODY SHAPE
Slender; tiny front legs, each bearing three toes, and no rear legs; bushy gills

NUMBER OF COSTAL GROOVES
29–37

BODY COLOR
Brown to black

PATTERN
Yellow stripes on back or sides

BODY SIZE
Adults up to 9 inches in total length

Description Dwarf sirens are small, eel-like salamanders with two legs and external gills. Each of the two front legs has three toes. Northern dwarf sirens are brownish to black with a yellow line on each side that runs from the snout to the tail. Many individuals lack sharply defined dorsal stripes, although they may have thin dark lines on the back. The Gulf hammock dwarf siren subspecies (*P. s. lustricolus*) has three narrow yellow stripes on the back and two on each side; the slender dwarf siren (*P. s. spheniscus*) has no stripes on the back and two stripes on each side. Southern dwarf sirens are light gray to brownish

top This southern dwarf siren from Putnam County, Florida, belongs to the subspecies known as the narrow-striped dwarf siren, which has five weakly defined stripes running the length of the body.

above The Gulf hammock dwarf siren subspecies has three narrow yellow stripes on the back and two on each side.

right Narrow-striped dwarf sirens have five stripes that are weakly defined or distinct but have irregular margins.

below The slender dwarf siren has no stripes on the back and two stripes on each side.

- ● northern dwarf siren
  Pseudobranchus striatus
- ● both northern dwarf and
  southern dwarf sirens
- ● southern dwarf siren
  Pseudobranchus axanthus

black with parallel yellow or tan stripes on the back and sides that run from the head to the tail tip. Stripe color and number vary among populations. Narrow-striped dwarf sirens (*P. a. axanthus*) have five stripes that are weakly defined or distinct but have irregular margins; the Everglades dwarf siren (*P. a. belli*) has five distinct stripes.

Larvae and Juveniles Hatchlings of both species have a fin on the back extending from the head to the tip of the tail. Hatchling northern dwarf sirens look like adults but are blacker and have an overall reddish hue. Southern dwarf siren hatchlings are brown with a yellowish stripe down the back and on each side.

Northern dwarf sirens have a dark body with a bold yellow line on each side that runs from the snout along the body and tail.

Similar Species The greater siren is similar in appearance but much larger (to 38 inches) and has 36–40 costal grooves and four toes on each foot; juveniles have a light stripe along the side of the body. The lesser siren (to 27 inches) has 31–35 costal grooves, and juveniles have red or yellow lines on the head. Amphiumas have two pairs of tiny limbs and lack external gills.

Distribution Northern dwarf sirens are found from the lower Coastal Plain of South Carolina south through Florida to Volusia County on the east and Hernando County on the west and in much of the Florida panhandle. Southern dwarf sirens occur only in peninsular Florida.

Habitat Dwarf sirens occupy cypress and gum ponds, swamps, ditches, marshes, limestone sinks, and other vegetation-choked freshwater wetlands. They prefer thick vegetation in the water column and decaying vegetation and mucky soils along the wetland margin. Mats of the exotic plant water hyacinth are frequently used microhabitats. Northern dwarf sirens are more often as-

Unlike members of the genus *Siren*, which have four toes per leg, dwarf sirens and other species in the genus *Pseudobranchus* only have three.

sociated with acidic cypress swamps. Southern dwarf sirens are more often associated with freshwater marshes and prairie ponds.

Behavior and Activity Dwarf sirens form cocoons in the mud when their wetlands begin to dry and can aestivate for several months encapsulated underground. Like other aestivating sirens, they absorb their gills and undergo changes in body mass and physiology during this time. They occasionally aestivate in S-shaped burrows as much as a foot below the surface of the substrate. Dwarf sirens occasionally vocalize with a faint, high-pitched squeak.

Reproduction Courtship and mating have not been described. Fertilization is likely external because males lack the ability to produce spermatophores. Females lay small numbers (probably fewer than 20) of eggs singly or in small clusters attached to vegetation in November to March. Clutch size is unknown. Eggs probably hatch in a few weeks, but individuals may take as long as 2 years to reach maturity.

Food and Feeding Dwarf sirens eat a wide array of small aquatic invertebrates, including many insects and their larvae, amphipods, ostracods, and worms. Their small mouth size limits the size and therefore the range of prey these salamanders can consume.

Predators and Defense Known predators include striped crayfish snakes, mud snakes, and southern watersnakes. Small alligators, turtles, wading birds, fish, and some predatory invertebrates undoubtedly eat dwarf sirens as well. Some adults emit an audible yelp when captured or prodded, a sound that might deter some predators. Because they are very slippery, their most effective defense if captured is to thrash about.

Conservation Too little is known about the population status of dwarf sirens anywhere to determine if numbers or populations have been lost or are declining. Undoubtedly, complete loss of wetlands and degradation of wetlands resulting from agricultural, residential, and silvicultural activities have caused population losses. The Gulf hammock dwarf siren has not been captured in Citrus and Levy Counties, Florida, since 1951 and may be extinct.

Taxonomy Three subspecies are recognized for the northern dwarf siren (*P. striatus*). The broad-striped dwarf siren (*P. s. striatus*) was described in 1824; the Gulf hammock dwarf siren (*P. s. lustricolus*) was described in 1951; and the slender dwarf siren (*P. s. spheniscus*) was described in 1949. The broad-striped dwarf siren is found from southern South Carolina south to the northeastern coastal flatwoods of Florida. The Gulf hammock dwarf siren is limited to Citrus and Levy Counties, Florida. The slender dwarf siren occurs from northwestern peninsular Florida to south-central Georgia and in much of the Florida panhandle. The southern dwarf siren (*P. axanthus*), described in 1942, has two recognized subspecies: the narrow-striped dwarf siren (*P. a. axanthus*) and the Everglades dwarf siren (*P. a. belli*), described in 1952. The narrow-striped dwarf siren occurs in the northern half of Florida, and the Everglades dwarf siren is found in the southern half of the state.

LESSER SIRENS

Lesser Siren *Siren intermedia*

Western Siren *Siren nettingi*

Seepage Siren *Siren sphagnicola*

BODY SHAPE	NUMBER OF COSTAL GROOVES	PATTERN	BODY SIZE
Eel-like and slender; feathery external gills; one pair of legs	31–35	Mostly uniform, but some have tiny black dots	Adults up to 27 inches in total length
	BODY COLOR Dark gray to black		

Description Lesser sirens are eel-like in body form and have external gills and only one pair of small front legs. Each of the two limbs has four toes. Adults are uniformly gray to olive green to nearly bluish black. Some individuals have tiny black dots on the back and sides. The belly is uniformly light gray. There are 31–35 costal grooves along the body between the front legs and the cloaca. The seepage siren is smaller than the other lesser sirens and has red gills.

top Lesser sirens can be more than 2 feet long. The gills are visible throughout the adult's life. Most individuals are grayish all over.

Lesser sirens have external gills and only one pair of small front legs.

Larvae and Juveniles Hatchlings have partially developed limbs and a dorsal fin that runs from the head to the tip of the tail. Hatchlings and small juveniles have a red band across the snout and along the side of the head but are otherwise uniformly brown to gray.

Similar Species Greater sirens are similar in appearance but are much larger as adults (to 38 inches) and have 36–40 costal grooves. Juvenile greater sirens have a light stripe along the side of the body and no red markings on the head. Dwarf sirens are small with several stripes and three toes on each limb. Amphiumas do not have external gills.

Distribution Lesser sirens occur from the Coastal Plain of Virginia south to the middle of peninsular Florida; west through south-central Georgia and Alabama through Louisiana; and north to southern Lake Michigan in Illinois, Indiana, and one small area in Michigan. The eastern lesser siren subspecies (Siren i. intermedia) is found from Virginia through Alabama. The western siren (S. nettingi) occurs from Mobile Bay, Alabama, westward across central Alabama

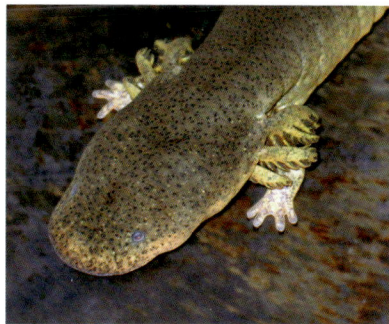

Some individuals have tiny black dots on the back and sides.

Adults are uniformly gray to olive green to nearly bluish black.

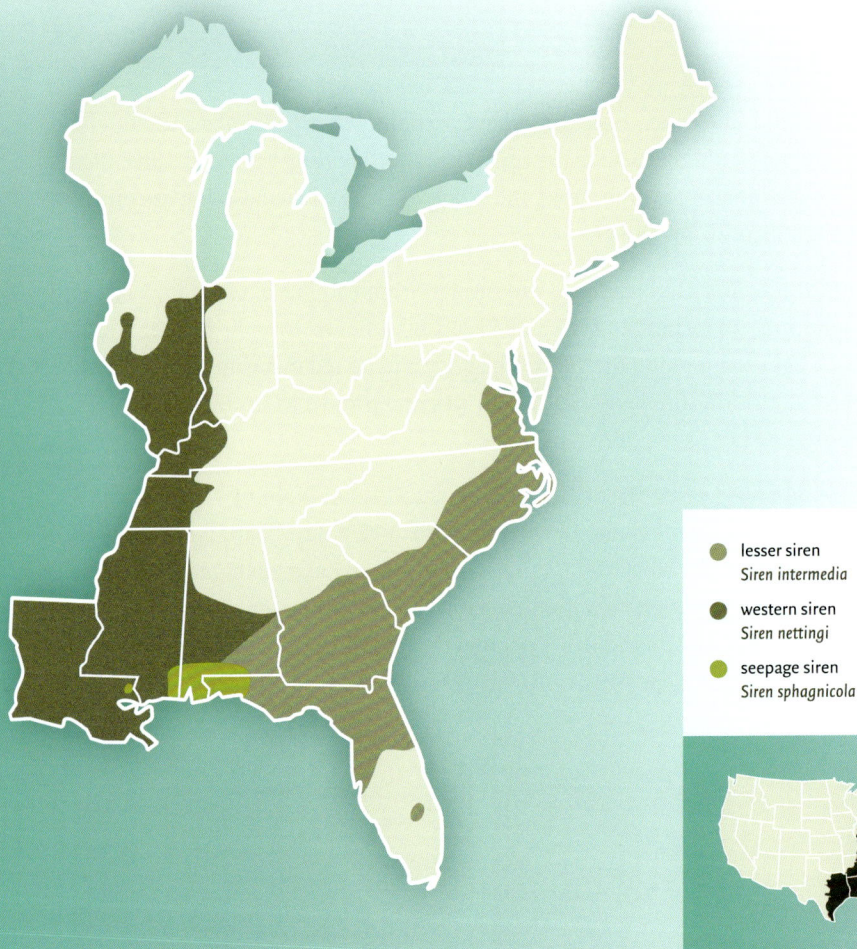

lesser siren
Siren intermedia

western siren
Siren nettingi

seepage siren
Siren sphagnicola

to eastern Texas and extreme northern Mexico. Its range extends north up the Mississippi River drainage to Lake Michigan. The seepage siren (S. *sphagnicola*) has a limited distribution, occuring mostly in the Florida panhandle from the tributaries of the Choctawhatchee Bay westward through the lower coastal plain of Alabama and Mississippi to the extreme eastern parishes of Louisiana.

Habitat Lesser sirens live in ditches, ponds, sloughs, backwaters of small sluggish streams, ponds, shallow lakes, marshes, and swamps where aquatic vegetation and organic sediment are plentiful. Hatchlings and juveniles prefer

The seepage siren (*S. sphagnicola*) is a smaller, red-gilled siren species.

vegetation-choked habitats and root masses. Lesser sirens can withstand water low in oxygen and acidity and can aestivate in the mud when their wetlands dry. Seepage sirens occupy shallow seeps, root mats, and sphagnum bogs.

Behavior and Activity Lesser sirens have both gills and lungs. They occasionally venture beyond the water's edge to hide under objects adjacent to water, and they may move considerable distances over land when rains create temporary waterways. Individuals aestivate in the mud by secreting a mucus cocoon around themselves and undergoing dramatic changes such as absorbing the gills, fat, and muscle, all of which regenerate when the pond refills. Activity slows in winter months, and lesser sirens in northern latitudes may hibernate in aggregations underground. Except when migrating to a nest or dispersing over land, these sirens remain within a small aquatic range less than 20 feet square. They feed at night. Lesser sirens sometime make a clicking or yelping noise that could be a communication with other sirens.

Reproduction Mating has not been described. Females lay 200–500 eggs in February and March in root masses and other dense vegetation. Larvae hatch in 1.5–2.5 months and remain in dense vegetation for protection from predators.

Food and Feeding Juveniles and adults eat a diverse array of prey, including insects and their larvae, worms, snails, freshwater clams, tadpoles, larval salamanders, and small fish. Egg cannibalism has been reported.

Predators and Defense Alligators, southern banded watersnakes, cotton-mouths, mud snakes, largemouth bass, and wading birds are known preda-tors. Feral hogs likely kill many sirens during droughts while the sirens are in cocoons in the mud. The tiny larvae and young sirens may be eaten by preda-ceous diving beetles, dragonfly nymphs, and other predatory insects. Nocturnal feeding may be a predator-avoidance strategy. One researcher noted shrill distress vocalizations.

Conservation Lesser sirens are apparently abundant in many areas, but pop-ulations suffer from aquatic pollution and habitat loss and fragmentation re-sulting from flood control programs and development. Feral hogs that root up aestivating salamanders in drying wetlands probably kill large numbers.

Taxonomy Two subspecies, both described in 1826, were recognized until quite recently: the eastern lesser siren (S. i. intermedia, with 31–33 costal grooves) and the western lesser siren (S. i. nettingi, with 35 costal grooves). In 2023, the western lesser siren was elevated to the species level as the western siren (S. nettingi). Also described in 2023, the seepage siren (S. sphagnicola) is a smaller, red-gilled siren species. Another yet-to-be-described siren is larger and has a different number of chromosomes than other siren species. Presumably all possess the same basic life histories and ecology described above.

GREATER SIREN   *Siren lacertina*

BODY SHAPE	NUMBER OF COSTAL GROOVES	PATTERN	BODY SIZE
Eel-like shape; large, bushy gills; one pair of legs	36–40	None or with small gold spots and fleck-ing	Adults up to 38 inches in total length

BODY COLOR
Olive, gray, or black

Description Greater sirens are large, eel-shaped salamanders with a pair of large, bushy gills; a shovel-shaped head; and one pair of front legs with four toes on each limb. The body is olive, dark brown, or nearly black with a gray belly. Individuals in some locations have well-defined gold flecking on the head and body. The lower sides are lighter than the back and often have faint greenish or yellowish dots or dashes. Some greater sirens are mottled with tiny yellowish dots on a gray or brown body. The belly has small greenish or yellowish flecks.

Larvae and Juveniles The larvae have fins on the back and tail, and both larvae and young juveniles are brown to gray with a yellowish stripe along each side from the gills to the base of the tail. The stripe fades within a year.

top Adult greater sirens can be more than 3 feet long. Sirens have two front legs but no hind legs.

Larvae and young juveniles are brown to gray with a yellowish stripe along each side from the gills to the base of the tail.

Similar Species Lesser sirens are similar in appearance but are smaller as adults (to 27 inches) and have fewer costal grooves (31–35 rather than 36–40). Juvenile lesser sirens lack the light stripe along the sides of the body. Dwarf sirens are small with stripes and three toes on each limb. Amphiumas have four tiny limbs and lack external gills.

Distribution Greater sirens occur from southern Maryland in the Coastal Plain south through peninsular Florida and west to southern Alabama.

Habitat Adults and juveniles occupy semipermanent to permanent aquatic habitats with abundant vegetation, such as streams, rivers, ditches, canals, freshwater marshes, Carolina bays, farm ponds, and lakes. Juveniles prefer shallow water but move into deeper water as they mature. Sirens use their shovel-shaped head to burrow in the bottom of rivers and wetlands.

Behavior and Activity Greater sirens are aquatic salamanders with functional gills. They may move short distances over land between bodies of water during rains, but probably must be submerged to move more than a few feet. Individuals may be active in all months of the year, but activity peaks during the early spring and summer months. Adults and juveniles are active at night and hide in burrows and under debris and litter on the bottom or in thick vegetation during the day. When wetlands dry, sirens secrete mucus that forms a cocoon around their body and aestivate until water returns, subsisting in the meantime on their

own body tissues. Laboratory studies indicate that large greater sirens can aestivate for more than a year under ideal conditions. Greater sirens can make clicking sounds when in proximity to others, but their purpose is unknown.

Reproduction Males apparently fertilize eggs as the females lay them, because no spermato-

Greater sirens are large and robust. Their size and 36–40 costal grooves distinguish them from lesser sirens.

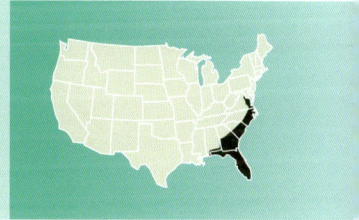

phores are produced. Little is known about courtship and mating. Females lay 100–500 eggs singly or in small groups attached to leaf litter in shallow water in February, March, and April. Hatching occurs after an incubation period of about 2 months. Females may remain with their eggs and guard them from predators.

Food and Feeding Greater sirens are generalists that feed on a wide variety of prey, particularly freshwater clams, snails, crayfish, and many types of insects and spiders; they also scavenge. They may be able to digest plant matter as well, because individuals are often found with large amounts of algae in their

Greater sirens are slippery and hard to hold. This photo captures a moment when the three individual sirens shown were not moving.

stomach. Adults occasionally eat aquatic vertebrates such as pond-breeding salamanders and small fish.

Predators and Defense Alligators, mud snakes, southern banded watersnakes, eastern green watersnakes, and wading birds are known predators. Feral hogs probably dig up sirens and eat them, especially during droughts when the salamanders are aestivating in the mud. Predatory insects and fishing spiders probably eat larvae and small juveniles. Several noises made by greater sirens have been suggested to deter some predators, especially birds and mammals, but confirmation that such soft sounds are effective is lacking. Sirens' primary defense is mucus secreted by skin cells that makes their body slippery and difficult to hold onto as they thrash and swim away.

Conservation Greater sirens appear to be relatively common in many parts of their range except in peripheral areas. The species is listed as Endangered in Maryland. Stream and river pollution may harm individuals and populations living in secondary waters. The effects of feral hog consumption of aestivating sirens may be severe. The conservation status of this aquatic species needs evaluation in all southeastern states.

Taxonomy The greater siren was first described in 1766. No subspecies are currently recognized, although one additional species from Alabama and northwestern Florida, the reticulated siren, has been described recently, and several more species are expected to be described in the *Siren* genus.

RETICULATED SIREN *Siren reticulata*

BODY SHAPE	NUMBER OF COSTAL GROVES	PATTERN	BODY SIZE
Eel-like shape; large external gills; only front legs	40	Dark, reticulated spotted pattern	Adults up to 24 inches in total length
	BODY COLOR Olive-gray with lighter sides		

Description The reticulated siren has an elongated, eel-like body, only two legs, no eyelids, a lateral line system, large feathery gills, and a horny beak instead of teeth at the front of the upper jaw. They have four toes on each forelimb, and three permanent gill branches with three associated gill slits. The body is olive-gray with lighter yellowish green sides. Dark spots form a conspicuous reticulated pattern from the gills to the tail, sometimes continuing down the sides toward the belly. The belly is a lighter olive green to yellowish color and sometimes is sparsely covered with irregular spots. The head is smaller and narrower in proportion to its overall size relative to the other siren species, and the tail is the longest of the sirens. Reticulated sirens have the highest costal groove count (38–42) of any siren.

top This species has remained secretive and elusive with limited localities to date.

The reticulated siren is one of the largest vertebrate species described in the past several years.

Larvae and Juveniles Little is known about the life history of this species. It is likely that they do not differ greatly from the other sirens in this regard.

Similar Species The only salamanders that the reticulated siren could be confused with are other sirens and amphiumas. The dwarf sirens are much smaller, have only one gill slit, and have three digits per foot. Lesser sirens have 31–38 costal grooves and are uniformly gray to olive green. The greater siren is olive to dark brown or gray with 36–40 costal grooves but no reticulated pattern. Amphiumas have tiny front and back legs and no external gills.

Distribution To date this species is only verified from three localities. It is known from the Alabama-Florida border at Lake Jackson (Covington County, Alabama, and Walton County, Florida), Eglin Air Force Base in Okaloosa County, Florida, and Fish River in Baldwin County, Alabama. This secretive, elusive salamander may be endemic to the Panhandle region of Florida and Alabama, and its geographic range potentially extends into southwestern Georgia.

Habitat The few specimens that have been found were from a heavily vegetated, shallow marsh (less than 3 feet deep); a beaver-impounded, clearwater stream with an associated bay swamp; and a blackwater stream with an associated bottomland forest. In at least one of these habitats the reticulated siren is sympatric with other sirens.

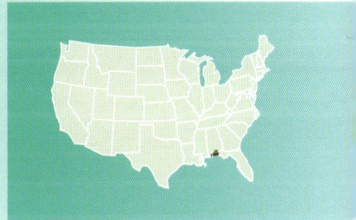

Behavior and Activity Little is known about the activity of reticulated sirens, but they are probably similar in behavior to lesser sirens and greater sirens.

Reproduction Mating has not been observed nor have eggs or nests. Reticulated sirens probably lay 200–500 eggs attached to leaf litter in shallow water in late winter to spring. Females likely remain with their eggs until hatching.

Food and Feeding Juveniles and adults probably prey on insects, insect larvae, worms, snails, clams, tadpoles, larval salamanders, and small fish.

Predators and Defense Predators probably include alligators, watersnakes, mud snakes, and wading birds like great blue herons. Sirens secrete mucus, making them slippery and hard to grasp, which allows them to elude some predators.

Conservation The restricted distribution and scarcity of observations of the basic life history and ecology of the reticulated siren make this species one of special concern. Currently, there is a lack of adequate data to assess its conservation needs. It is found within a hotspot of ecological biodiversity that includes vulnerable wetlands and the threatened longleaf pine ecosystem.

Taxonomy Decades passed between when the first individuals were collected in 1970 and when the species was formally described in 2018. Recent research indicates that the reticulated siren is not a close relative of the other species of sirens and that other siren species remain undescribed.

aquatic cave salamanders

GEORGIA BLIND SALAMANDER *Eurycea wallacei*

BODY SHAPE
Slender; long head;
tail flattened from
side to side

**NUMBER
OF COSTAL
GROOVES**
12 or 13

BODY COLOR
Pink and white

PATTERN
None

BODY SIZE
Adults up to 3 inches
in total length

Description These unusual little salamanders have a large, slightly flattened head; a squared-off snout; long, bright red external gills; four long, spindly legs; and a flattened tail with fins. Adults retain all of the typical larval characteristics. The degenerate eyes are dark spots buried below the skin. The body is pink to silvery white and translucent, and may have traces of dark specks or blotches.

Larvae and Juveniles The larvae have not been described because they have never been seen by scientists. Juveniles resemble adults in most characters. Small individuals have some wdark pigment. They do not metamorphose.

Similar Species The other troglobitic salamanders with which this species might be confused, such as the Tennessee cave salamander and grotto salamander, occur well outside the range of the Georgia blind salamander.

top The Georgia blind salamander has a large, flat head; squared-off snout; long external gills; and long spindly legs. The tiny eyes are dark spots, and the body is translucent.

Distribution Georgia blind salamanders are found in the Coastal Plain of southwestern Georgia and in several counties in the Florida panhandle. The only known localities are in caves that provide access to the Floridan Aquifer.

Habitat People have found Georgia blind salamanders in limestone caves, sinkholes, and artesian springs, some with deep silt layers. These completely aquatic salamanders occupy subterranean streams and pools, especially in areas where bat guano provides nutrients to the aquatic system.

Young Georgia blind salamanders (top) have more dark pigmentation than adults (bottom).

Behavior and Activity Georgia blind salamanders are very sensitive to light and movements in the water. They will swim quickly in an upward spiral when disturbed but then settle back down on the substrate and remain motionless. Their seasonal activity period is unknown but may be regulated by stream flows and the timing of nutrient input into the caves.

These subterranean salamanders are true paedomorphs that retain all the larval characteristics as adults. Paedomorphosis is a common trait of true troglobitic animals.

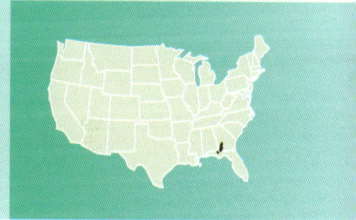

Reproduction These small salamanders reach maturity at a total length of about 2 inches. Females carrying enlarged eggs have been found in May and November. Nothing is known about mating behavior, clutch size, egg-laying dates, or other aspects of reproduction.

Food and Feeding Georgia blind salamanders either bump into potential prey items or sense their presence with the lateral line system, a row of sensory cells on the sides that detect vibrations in the water. Adults and juveniles eat small isopods, amphipods, copepods, mites, and small beetles.

Predators and Defense Georgia blind salamanders co-occur with the Dougherty Plain blind crayfish, which may be a primary predator. Several species of fish are known from the same underground water system, and they, too, could easily eat these salamanders. Predation has never been observed in nature.

Conservation This unique salamander is threatened by any human activity that affects caves and aquifers, including chemicals associated with agriculture, urbanization, and industrial activities. Loss of aquifer water from drawdowns for agriculture and drinking water for cities and towns is a potentially severe threat. The species is listed as Threatened in Georgia.

Taxonomy The Georgia blind salamander was first known to science when specimens found in an artesian well were described in 1939. No subspecies are recognized. This species was formerly placed in its own genus, *Haideotriton*.

TENNESSEE CAVE SALAMANDERS

Berry Cave Salamander   *Gyrinophilus gulolineatus*

Tennessee Cave Salamander   *Gyrinophilus palleucus*

BODY SHAPE	NUMBER OF COSTAL GROOVES	BODY COLOR	BODY SIZE
Slender; large, spatulate head; gills present; small eyes	17–19	Pink to salmon to brownish purple	Adults up to 8 or 9 inches in total length
		PATTERN Uniformly colored or with dark spots	

Description   Tennessee cave salamanders have a slender body; a large head with a broad, squared-off snout; small degenerative eyes; long external gills; and an oarlike tail. Gills are present throughout life. Body color varies among populations from pink, to light tan without dark spots, to brown, to brownish purple with abundant black spots (*see* Taxonomy). Berry Cave salamanders are similar to Tennessee cave salamanders in body form, size, and color but have a more spatula-shaped snout, and some have a dark stripe or patch under the throat.

top   Tennessee cave salamanders spend their entire lives in water in caves and are true troglobites.

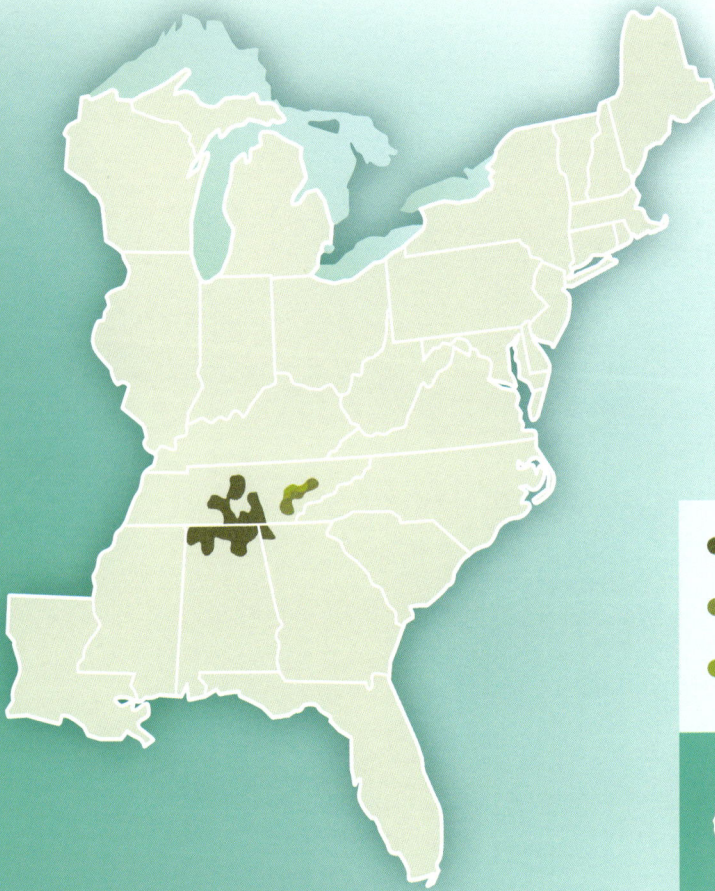

- ● Tennessee cave salamander
  Gyrinophilus palleucus
- ● both Tennessee cave
  and Berry Cave salamanders
- ● Berry Cave salamander
  Gyrinophilus gulolineatus

Juvenile Tennessee cave salamanders have small external gills.

Larvae and Juveniles Hatchlings have not been described. Juveniles resemble the adults in the same cave system.

Similar Species Larvae of the cave salamander (*Eurycea lucifuga*) are generally uniformly pigmented with three rows of light spots down each side of the body, and they have a shorter snout. Spring salamanders lack external gills as adults. The gilled larvae of spring salaman-

right Except for a few individuals like this one, Berry Cave salamanders retain their external gills throughout life. This individual has the species' distinctive oarlike tail and spatula-shaped snout.

below This is a typical Tennessee cave salamander adult with external gills. The oarlike tail is prominent in this individual.

ders resemble the Tennessee cave salamander in body form but have larger eyes. Adult spring salamanders are salmon to pinkish with black spots and streaks; the larvae are yellowish brown to gray with black flecking. The eye diameter of Tennessee cave salamanders is less than 25 percent of the length of the snout but is more than 30 percent of the snout length in spring salamanders.

Distribution These two salamander species are restricted to cave systems in central and eastern Tennessee, northwestern Georgia, and northern Alabama. Tennessee cave salamanders are found at the southeastern edge of the Cumberland Plateau in central Tennessee, northern Alabama, and northwestern Georgia. Berry Cave salamanders inhabit caves in the Valley and Ridge physiographic province in Knox, McMinn, Meigs, and Roane Counties, Tennessee.

Habitat These obligate aquatic salamanders spend their entire lives in water inside caves, in shallow pools with rocky and sandy bottoms. They are paedomorphic and are true troglobites. Berry Cave salamanders have been found in pools with clay or mud substrates.

Behavior and Activity Little is known about the behavior and ecology of these salamanders. Some individuals appear to be sedentary and rarely move from the pool in which they are first discovered; others have been known to move more than 400 feet within a stream. Winter and spring floods may wash some salamanders downstream, but individuals apparently seek cover under rocks or in crevices to avoid being washed away. Although these salamanders usually retain larval characters throughout their lives, laboratory experiments demonstrate that they can metamorphose, losing gills and tail fins and enlarging the eyes to normal size. A few naturally metamorphosed individuals have been found.

top All populations of Tennessee cave salamanders, like this Big Mouth Cave salamander, are threatened by the pollution of caves and other disturbances from agriculture and urbanization.

bottom Tennessee cave salamanders can be up to 8 or 9 inches in total length.

Reproduction Mating appears to occur in summer. Females may skip reproduction for one or more years. Eggs have not been described. Larvae have been found in December and February.

Food and Feeding Adults and larvae eat worms, amphipods, isopods, crayfish, beetles, fly larvae, stoneflies, mayflies, caddisflies, and small crustaceans. Cannibalism by larger individuals on smaller ones has also been reported.

Predators and Defense Two predators have been identified: an American bullfrog inhabiting a cave, and other Tennessee cave salamanders (cannibalism). Crayfish may also eat the larvae and small adults. The salamanders readily seek refuge beneath rocks when disturbed.

Conservation Cave salamanders are threatened primarily by pollution and silt entering the cave system as a result of land disturbance (runoff from agriculture, septic tanks, roads, deforestation, and urbanization). Cave systems in Tennessee are threatened as the expanding human population clears land and places houses within karst watersheds. Both species are listed as Threatened in Georgia and Tennessee and are considered Species of High Conservation Concern in Alabama. Studies of how the impacts to caves from outside sources affect these unusual salamanders will help to focus habitat management efforts that will in large part determine their fate.

Taxonomy Two subspecies of Tennessee cave salamanders are recognized: the pinkish cave salamander (G. p. palleucus), described initially in 1954; and the spotted Big Mouth Cave salamander (G. p. necturoides), described in 1962. The Berry Cave salamander, described in 1965, was once considered a subspecies of the Tennessee cave salamander.

WEST VIRGINIA SPRING SALAMANDER

Gyrinophilus subterraneus

BODY SHAPE
Large; usually with gills; small eyes

NUMBER OF COSTAL GROOVES
17

BODY COLOR
Brownish pink to pale pink

PATTERN
Light gray reticulation on body with a series of pale-yellow spots on sides

BODY SIZE
Adults up to 7 inches in total length; larvae up to 4.4 inches from snout to vent

Description The West Virginia spring salamander is a large, troglobitic, robust relative of the spring salamander (*Gyrinophilus porphyriticus*) with pale skin, dark reticulations, more teeth at the front of the upper and lower jaw, and reduced eye size. The head is wide, and there are two or three irregular rows of pale-yellow spots down each side. The characteristic canthus rostralis (a distinct line from the eye to the nasolabial groove) of the sympatric spring salamander is indistinct.

Larvae and Juveniles Larvae, the dominant life stage that has been observed, reach a large size and are pale with large, feathery, red gills. Large larvae appear to be sexually mature indicating the possibility of neoteny or paedomorphosis.

top A transformed, adult West Virginia spring salamander from General Davis Cave in West Virginia

Metamorphosis to adults takes place at snout-to-vent lengths greater than 3.75 inches. Most transformed adults appear emaciated in appearance, not nearly as robust as late-stage larvae. It has been hypothesized that the large, late maturing larvae might be the result of either insufficient terrestrial resources and/or the absence of predators in the aquatic cave habitat.

Similar Species The spring salamander (*G. porphyriticus*) is similar in appearance but brighter in coloration (red to salmon pink) and has larger eyes, fewer teeth, and a canthus rostralis. The other two species in this genus have restricted ranges and never overlap spatially with the West Virginia spring salamander.

Distribution This microendemic salamander is found only in the cave region of the Upper Kanawha River basin in West Virginia, specifically the General Davis Cave in Greenbrier County. It has been found as deep as 0.6 miles into the cave.

Habitat Adults have been collected streamside in the cave in accumulated leaf pack; larvae live in the stream proper. Most larvae are on the stream bottom, but some climb up the stream walls. Large larvae inhabit the shallow pools and avoid cover objects and riffle zones.

Behavior and Activity The life history of this secretive and rare species is relatively unknown.

Reproduction Nothing is known about the reproduction of West Virginia spring salamanders, but it is probably similar to reproduction in the spring salamander. Females seem to lay eggs away from the main cave stream in small, protected, feeder springs. This is probably where very young larvae hide until they reach sufficient size to avoid most predators

Most specimens found of the endangered West Virginia spring salamander are in the larval stage.

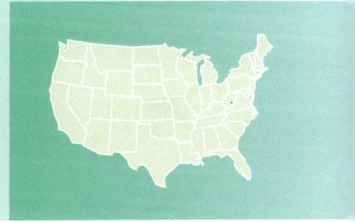

Food and Feeding The small eyes and cave habitat of the West Virginia spring salamander make capturing food dependent upon nonvisual senses, such as the tactile and olfactory senses, that allow them to detect movement and pick up chemical cues. They are opportunistic salamanders, consuming worms, amphipods, isopods, ostracods, copepods, and other common invertebrates. Detritus is often found associated with food items. Large larvae may occasionally eat small salamanders. Most observed specimens of West Virginia cave salamanders have been robust and large, indicating successful foraging even in total darkness.

Predators and Defense The cave environment provides protection from most predators, and no specific defense strategies have been observed except for movement away from disturbances. The large larvae are usually out in the open and not under cover, suggesting a lack of predators in the cave habitat.

Conservation The West Virginia spring salamander represents a very small, isolated cave population that is susceptible to decline and therefore listed by the International Union for Conservation of Nature as Endangered. The population appears stable in numbers, but the population total is probably less than 250. It is known from only one cave (General Davis Cave) out of more than 4,100 found in West Virginia. The cave itself, protected by an easement by the Nature Conservancy, serves as a hibernaculum for the endangered Indiana bat and is home to several other cave-dwelling animals. Although the cave entrance is owned by the Nature Conservancy, the principal source of the cave stream and the entire watershed feeding this stream (Muddy Creek) remains unprotected, making it vulnerable to any significant land use changes. The population is fragile and vulnerable to siltation, land clearing/logging, degradation, pollution, disease, and overcollection. West Virginia protects the salamander as a Species of Concern. The U.S. Fish and Wildlife Service has a listing review plan to determine whether or not to list this species under the U.S. Endangered Species Act.

Taxonomy Since the species was first described in 1977, the taxonomic validity of *G. subterraneus* has been questioned by some researchers. Some key arguments to support species recognition are morphological differences (larger larval size, decreased eye size, and increased number of teeth) between the West Virginia spring salamander and the sympatric, local populations of spring salamander (*G. porphyriticus*) and the lack of evidence of hybridization between the two forms when found together.

stream and seep
salamanders

STREAMSIDE SALAMANDER *Ambystoma barbouri*

BODY SHAPE
Robust; small,
rounded head

**NUMBER
OF COSTAL
GROOVES**
14 or 15

BODY COLOR
Grayish brown to
black

PATTERN
Tan to gray lichen-
like markings

BODY SIZE
Metamorphs 2
inches; adults up to
5.5 inches in total
length

Description Most adult streamside salamanders are grayish brown to black
with a dense pattern of tan to light gray markings that resemble lichens. Some
adults lack the pattern on the back and are uniform in color. The belly is gray to
grayish black with light grayish blotches. The snout is short, wide, and rounded.

Larvae and Juveniles The aquatic larvae are uniformly dark greenish brown
to olive and have feathery gills. The dorsal tail fin extends forward to behind
the head. Younger larvae have three to six pairs of light blotches or saddles on
the back and tail, but older larvae do not. A light-colored line of small spots
may run along each side of the body. The belly is cream, and the tail fin may
have greenish brown mottling. Larvae in ponds have a broad tail and large,
bushy gills; larvae in streams are slim with a narrow tail and small gills. Larvae

top A mature adult streamside salamander of the uniform grayish brown phase. Note the short
snout and robust body.

above An adult streamside salamander with a gray body and light gray lichen-like markings.

left The large, bushy gills and broad tail fins indicate that this streamside salamander larva is from a pond. Slender larvae with small gills are typical of stream habitats.

transform into terrestrial juveniles in 1–3 months depending on temperature. Newly metamorphosed juveniles are uniformly brown to brownish gray with little or no pattern.

Similar Species The streamside salamander closely resembles the small-mouthed salamander, but geographic location can often be used to distinguish the two species. Small-mouthed salamander adults are brownish gray to black with clusters of light specks or a lichen-like pattern mostly concentrated on the sides. Juvenile small-mouthed salamanders closely resemble streamside salamander juveniles but develop the specks and lichen-like pattern within 1–2 months after metamorphosis. The teeth of the upper and lower jaws of both species occur in multiple rows; however, in streamside salamanders the inner cusp of the teeth of the upper jaw is short, rounded, and spadelike, and in small-mouthed salamanders the teeth of the upper jaw have long, pointed inner (lingual) cusps. Marbled salamander juveniles are brown to black with tiny gray to silver specks that may occur as a lichen-like pattern on the back. Spotted salamander juveniles are olive to gray with fine gray specks on the back but not on the tail. Mole salamanders are short and chunky, have a large head, and have no markings or light gray flecking. Juvenile Jefferson salamanders are uniformly dark gray, occasionally with brownish yellow specks on the sides.

Distribution Streamside salamanders are found in north-central Kentucky, southwestern Ohio, and southeastern Indiana, with isolated populations in western Kentucky, central Tennessee, and southwestern West Virginia.

Habitat Outside the breeding season, adults remain underground in hardwood forests adjacent to streams and ponds. Juveniles occupy similar habitat as well as upland forested areas. Breeding sites include small, fish-free headwater streams with alternating pools and riffles with large, flat limestone rocks, and a wide variety of woodland pools and pondlike wetlands.

STREAMSIDE SALAMANDER *Ambystoma barbouri*

A subadult streamside salamander with profuse light gray, lichen-like markings. The markings fade with age, and mature adults are uniformly colored with few markings.

Behavior and Activity Adults emerge on rainy nights in late October through early March and migrate to streams and pools for mating. The females lay eggs in the water between early December and early April, then migrate back to their upland burrows. Eggs hatch from mid-March through May, and larvae remain in the aquatic habitat until they transform into juveniles 1–3 months later.

Reproduction This is the only mole salamander (family Ambystomatidae) that breeds in streams. Males arrive at breeding sites in pools or flowing sections of limestone streams before females. After the females arrive, small groups of both sexes court and mate beneath large rocks over a period of several months, then leave to return to their terrestrial burrows in late winter to early spring.

> **DID YOU KNOW?**
>
> *All salamanders lay eggs with the exception of some of the fire salamanders of Europe that are livebearers.*

Females lay up to 260 eggs in a single layer on the underside of rocks, logs, or other submerged or partially submerged cover objects from late December to early April; several females may use the same rock. Some populations of streamside salamanders breed in small woodland pools, ditches, and flooded fields; and females attach eggs to logs, rocks, or grass stems.

Food and Feeding The aquatic larvae eat small invertebrates and zooplankton, especially aquatic isopods. Larval densities are sometimes so high that the young salamanders eat most of the prey in their pool. Cannibalism occurs occasionally. Both juveniles and adults are opportunistic feeders that eat most types of invertebrates they encounter in the leaf litter of the forest floor.

Predators and Defense The primary predators, especially of the larvae, are crayfish and green sunfish, although streamside salamanders do not usually occupy portions of streams where fish occur. Other predators include garter snakes and northern watersnakes. A type of flatworm has been known to prey on smaller individuals. Adults use biting, immobility, head hiding, and body posturing to avoid predation. Tail waving attracts predators to the part of the body where noxious compounds are secreted. Larvae hide under leaf litter and rocks in the stream bottom. Streamside salamanders are often active out in the open during daylight hours.

Conservation The Tennessee Wildlife Resources Agency lists this species as Deemed in Need of Management and as an s2 species—a Natural Heritage ranking indicating that the species is imperiled because of its rarity. Conversion of forests and fields into housing developments threatens middle Tennessee populations. Stream pollution and habitat loss and alteration from deforestation threaten local populations throughout the range.

Taxonomy The streamside salamander was first described in 1989. No subspecies are recognized.

SEEPAGE SALAMANDER *Desmognathus aeneus*

BODY SHAPE
Tiny and slender

NUMBER OF COSTAL GROOVES
13 or 14

BODY COLOR
Brown

PATTERN
Usually a single light stripe with dark borders down the back

BODY SIZE
Adults up to 2.5 inches in total length

Description These tiny, slender salamanders have a round tail that makes up less than half the total length. A pale yellow, tan, or reddish brown, slightly wavy to straight-edged stripe usually bordered by dark brown occurs on the back and tail, although some individuals have a series of chevrons rather than a distinct stripe. A thin, dark, Y-shaped patch behind the eyes connects to a dark line or series of small herringbone-like lines that may continue into the stripe along the middle of the back. A light stripe runs from the eye to the jaw. The belly is uniform or mottled with brown and white.

Larvae and Juveniles Hatchlings have a proportionally larger head and shorter tail than adults. Juveniles are patterned as adults, but their coloration is brighter. There is no aquatic larval stage.

top The Y-shaped marking behind the eyes of this seepage salamander connects to the stripe along the middle of the back.

Similar Species The top of the head is smooth in the seepage salamander but rugose in the pygmy salamander. The stripe on the back of Ocoee salamanders forms a zigzag pattern, and this species lacks a Y-shaped mark on the head. Southern red-backed salamanders lack the eye-to-jaw stripe and the Y-shaped mark.

Distribution Seepage salamanders are found in the southern Appalachian Mountains from southeastern Tennessee and southwestern North Carolina through northern Georgia into Alabama. Disjunct populations occur in the Fall Line hills of central Alabama and the Piedmont of Georgia and in Oconee County, South Carolina.

Habitat These small salamanders are strongly associated with seeps and moist areas around small streams. They live between the moist leaf litter and the ground but may also be found under logs and mats of mosses. Although the highest population densities occur in seepage areas in cove forests, individuals occasionally move out into hardwood forests.

above This female seepage salamander represents one of the smallest species of salamander in North America north of Mexico.

right Adult seepage salamanders are characterized by a broad brown stripe along the back and tail bordered on each side by a dark stripe.

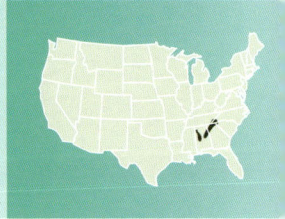

An adult female
with her eggs

Behavior and Activity Adults and juveniles remain hidden under leaf litter most of their lives. They do not aestivate and do not migrate.

Reproduction Mating is terrestrial and occurs in February and March and from late July to October. After an elaborate courtship, the female picks up a spermatophore the male has deposited on the substrate. Females lay 3–19 eggs in moss or under decomposing logs in or near seeps, and remain with them until they hatch 6–9 weeks later. Eggs have been found from February to late May and hatch in late May to August.

Food and Feeding These diminutive salamanders forage under leaf litter, feeding on small invertebrates such as isopods, amphipods, earthworms, beetles, mites, and spiders. The observation of an adult male eating a juvenile indicates that cannibalism occurs.

Predators and Defense Natural predators are unknown, but ringneck snakes, spring salamanders and other predatory salamanders, and leaf litter–foraging birds are likely to eat these little salamanders. Seepage salamanders of all sizes become immobile when they are uncovered, which probably makes them inconspicuous to predators.

Conservation Populations are vulnerable to loss of mountain headwater streams and their associated seepages, but the seepage salamander is not protected in any state in the Southeast. Clearing of hardwood forests around headwater streams alters the seepage environment by eliminating leaf litter and moisture.

Taxonomy The seepage salamander was first described in 1947. No subspecies are recognized. Recent research has revealed that this species may be a species complex, and it may be divided into multiple species in the future.

SOUTHERN DUSKY SALAMANDERS

Southern Dusky Salamander   *Desmognathus auriculatus*

Pascagoula Dusky Salamander   *Desmognathus pascagoula*

Valentine's Southern Dusky Salamander   *Desmognathus valentinei*

Carolina Swamp Dusky Salamander   *Desmognathus valtos*

BODY SHAPE	**BODY COLOR**	**PATTERN**	**BODY SIZE**
Stout; medium-sized	Dark brown to gray-black	One or two rows of white spots on each side	Adults up to 6 inches in total length
NUMBER OF COSTAL GROOVES			
14			

Description Southern dusky salamanders (*D. auriculatus*) have a keeled tail that is flattened side to side and becomes blade-like toward the tip, and a light eye-to-jaw stripe that is yellowish to orange-red. The back is dark brown to black with one or two rows of small white to reddish spots on each side between the front and hind limbs. Fine white specks may dot the back. Some individuals in

top A keeled tail flattened from side to side and orangish stripe behind the eye are characters of the southern dusky salamander.

peninsular Florida have a reddish back suggesting a light stripe. The belly is dark brown to black and sprinkled with small, distinctive white spots. Males are typically larger than females, possess a mental gland on the tip of the lower jaw, have a proportionally longer ventral opening, and can, as they age, have larger "jaw" muscles giving them a jowly appearance. The dorsal pattern of the adult Valentine's southern dusky salamander (D. valentinei) is nondescript, lacking defined spots or edges. This species often has a reddish-brown coloration on the anterior half of the tail. Their belly is mottled dark brown and tan with few white flecks. A sinuate groove strongly indents the skin running from the back of the eye above the light-colored cheek patch to the throat fold. Below this groove the skin is lighter gray with six or seven tiny, faint light gray spots. The Pascagoula dusky salamander (D. pascagoula) is smaller and has a more defined dorsal pattern with irregular whitish portholes-like spots that are distinct from the whitish flecks. The white spots occur in as many as three rows on the sides and tail. Unlike the Valentine's southern dusky salamander, the Pascagoula dusky salamander usually has a brighter orange to yellowish-orange stripe from the back of the eye to the angle of the jaw.

Larvae and Juveniles Larvae are black to dusky brown with pale spots on the back, and their gills are dark and exceptionally bushy. The larvae in the Florida panhandle may have tiny white spots on the sides of the body and tail. Juveniles have six or seven pairs of light spots on the back. Three rows of tiny spots mark-

right Larvae are black to dusky brown with pale spots.

below A recently metamorphosed juvenile

Carolina swamp dusky salamander
Desmognathus valtos

southern dusky salamander
Desmognathus auriculatus

Pascagoula dusky salamand
Desmognathus pascagoula

Valentine's southern
dusky salamander
Desmognathus valentinei

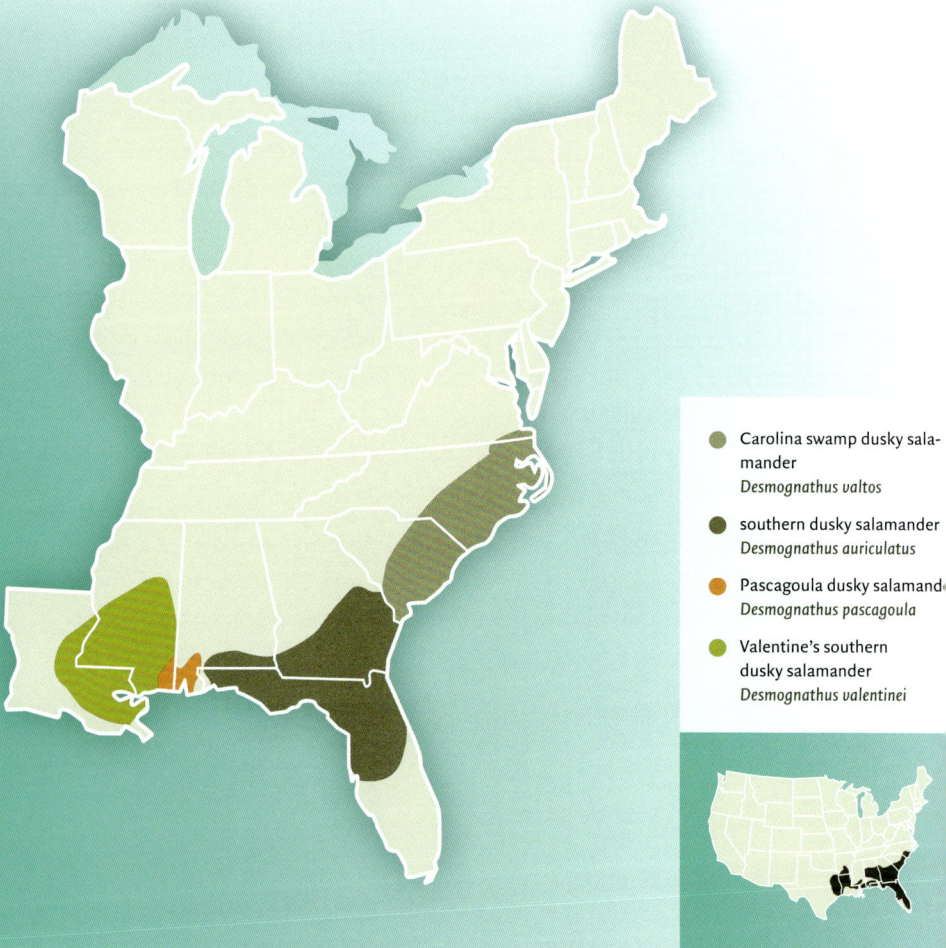

ing the presence of the lateral line system are visible in larval Valentine's dusky salamanders, but the typical line from the eye to the jaw is not clearly visible.

Similar Species Spotted dusky salamanders have a cream to tan belly with ir-regular dark brown mottling; otherwise, they are similar in color and pattern. Northern dusky salamanders often have a light, often reddish, irregular stripe on the back that extends onto the tail.

Distribution Southern dusky salamanders are found in the Coastal Plain physiographic province from southeastern Virginia south to central Florida

above The dorsal pattern of the adult Valentine's southern dusky salamander lacks defined spots or edges.

right This species often has a reddish-brown coloration on the anterior half of the tail.

and west to Texas. Valentine's southern dusky salamander has been found from extreme southwestern Alabama to southern and central southeastern Mississippi to eastern Louisiana. This includes the lower Mississippi, Big Black, Amite, Bogue Chitto, Pearl, and Pascagoula River drainages. Its western delineation and eastern limits still need to be refined. The Pascagoula dusky salamander has a small range in the Gulf Coastal Plain of southeastern Mississippi and the southwestern edge of Alabama. It is currently restricted to the lower Pascagoula, Escatawpa, and Mobile River drainages. Historical records indicate that they may have once inhabited the northeastern side of the Mobile-Tensaw River delta. The redescribed southern dusky salamander (D. *auriculatus*) is found from mid-peninsular Florida north to east-central Georgia and southern Georgia to extreme southern Alabama and all but the western edge of the Florida panhandle. Those populations northeast of Georgia in

This southern dusky salamander found shelter from predators inside a dead snail shell.

the Atlantic Coastal Plain currently remain designated as *D. auriculatus*, but several researchers indicate those genetically divergent populations will be classified as distinct species in the future. The drastic decline and local extirpation of many of these populations make taxonomic revisions difficult.

Habitat These very aquatic salamanders are found in muddy wetlands in southern floodplain habitats and in swamps, seeps, sloughs, small backwater streams, and cypress ponds. They live in deep, soupy, mucky organic substrates and can also be found under logs and leaf litter in seepage areas and mucky, acidic areas. They are not normally found in first- or second-order streams. They retreat beneath the peat layers when water levels are low.

Behavior and Activity The seasonal activity cycle is unknown, but these dusky salamanders are likely to be active in their microhabitats as long as the surface water is not frozen. They tend to move around very little except during droughts and cold temperatures, when they burrow deeper into the muck and peat. They are seldom found away from seeps and mucky areas but will enter more terrestrial habitats to forage on rainy nights. Adults may defend territories as other salamanders in this group do, but that behavior has not been documented.

Reproduction The little that is known about mating in the southern dusky salamander indicates that it occurs in late summer in shallow water or muck. Females lay 9–26 eggs beneath objects on the surface, in sphagnum moss, and in cypress logs less than 6 feet from water in September to December. Eggs are deposited in small, compact clusters and cling to one another. Females attend the eggs until hatching. Larvae hatch in October to January, and metamorphosis takes place in January and February.

Food and Feeding Adults and juveniles eat beetles and their larvae, fly larvae, spiders, and worms. The larvae undoubtedly eat small crustaceans and other aquatic invertebrates.

Predators and Defense Watersnakes and redfin pickerels are known to eat adults. Feral pigs rooting for food in drying wetlands probably eat them as well.

The Pascagoula dusky salamander has a small range in the Gulf Coastal Plain of southeastern Mississippi and southwestern Alabama.

Adults are cannibalistic and will eat eggs and juveniles. When exposed, these salamanders will try to escape quickly into muck, peat, and burrows.

Conservation Many southern dusky salamander populations have declined dramatically or disappeared in Florida, Georgia, and Alabama since the 1970s. This decline has been severe and has occurred in many seemingly pristine habitats, leading researchers to suspect that an unknown pathogen may be involved. Loss of headwater and floodplain wetlands is a primary threat, in addition to pollution and siltation that damage water quality. Feral pigs constitute an underappreciated threat to populations that needs further evaluation. The reason why declines have occurred in some parts of the range but not others also needs to be determined. Regulations restricting collection for personal use—such as for fish bait—are the only protection this species receives in any state.

Taxonomy Recent molecular studies indicate that the southern dusky salamanders may comprise as many as three additional species. Future work may add several new species in the Atlantic Coastal Plain from the Savannah River on the Georgia–South Carolina border to southern Virginia. These populations have been initially shown to be genetically divergent from D. auriculatus, as redescribed in 2017. The species was first described to science in 1838 and was long considered to be a subspecies of the northern dusky salamander. Valentine's southern dusky salamander was described in 2017, and the Pascagoula southern dusky salamander was described in 2022.

An additional southern dusky salamander from the Atlantic Coastal Plain was described in 2022. This newly described species has been named the Carolina swamp dusky salamander (Desmognathus valtos). Its proposed range is from Liberty County, Georgia, through South Carolina to Beaufort County, North Carolina.

IMITATOR SALAMANDER   *Desmognathus imitator*

BODY SHAPE
Small with round
tail

**NUMBER
OF COSTAL
GROOVES**
14

BODY COLOR
Gray, brown, black

PATTERN
Colorful cheeks;
wavy-edged stripe
on back

BODY SIZE
Adults up to 4 inches
in total length

Description The imitator salamander is related to the mountain dusky group but is unique in having yellow, orange, or red cheeks in addition to the light (but sometimes dark) line between the eye and the angle of the jaw. The brown to black body and tail usually have a bold, strongly wavy-edged, light-colored stripe. Some old individuals lack both the colored cheek patches and the stripe and are a uniform grayish black. The belly of adults is uniformly gray. The tail, as in mountain duskies, is round at the base and lacks a keel.

Larvae and Juveniles The juveniles are similar to adults in color and pattern. The larvae have not been described.

Similar Species The Ocoee salamander is most likely to be confused with the imitator. Those occurring within the range of the imitator salamander do not

top The orange cheek and the line between the eye and back of the jaw combine to form the large, colorful blotch characteristic of many imitator salamanders.

have colored cheeks, but they might be mistaken for older imitators that lack the colored cheek patches. If the stripe on the back is present, the imitator generally has a wavy-edged pattern and the Ocoee typically has a relatively straight-edged pattern. Imitator salamanders are not as abundant at high elevations as Ocoee salamanders are.

Distribution Imitator salamanders are found in a small area in the Great Smoky, Plott Balsam, and Balsam Mountains in eastern Tennessee and southwestern North Carolina.

Habitat Imitator salamanders can be found on and under wet rocks at the edge of streams; in seepage areas; on wet rock faces; and in the leaf litter alongside headwater streams in spruce, fir, and hardwood forests at elevations above 2,200 feet.

Behavior and Activity Imitator salamanders have been found only in warm-season months, but their habitats in the high mountains are very difficult for biologists to reach in winter. During the warm season the salamanders are nocturnal and seek cover during daytime. Adults disperse into nearby forested habitat during wet periods.

Reproduction Mating probably occurs in spring and fall but has not been observed in nature. The male deposits a spermatophore on the moist substrate that the female picks up with her cloaca following an elaborate courtship. Females

Look for these imitator salamanders at the edges of streams, in seepages, and on wet rock faces. Not all Imitator salamanders have the red-orange cheek patch.

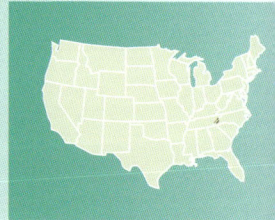

lay 13–30 eggs, attaching them to the underside of rocks and placing them in moss in moist areas alongside headwater streams. Nothing is known about the larval stage, but it may be similar to that of the Ocoee salamander and may last less than 1 year.

Food and Feeding Imitator salamanders feed at night from rocky perches, and possibly in the adjacent leaf litter as well. The diet has not been investigated but almost certainly includes a wide array of small invertebrates such as insects, worms, and spiders.

Little is known about the life history and ecology of this salamander, but advances in snow vehicle technology might allow scientists to study this species more completely.

Predators and Defense No predators have been documented, but ringneck snakes, garter snakes, watersnakes, and spring salamanders are probably among them. When discovered, the salamander either flees or becomes immobile. The red cheek color may mimic that of the red-cheeked salamander, a species that visual predators consider distasteful.

Conservation This high-elevation salamander may be subject to habitat and thermal changes associated with predicted climate change. Most populations are within Great Smoky Mountains National Park and are thus afforded some protection against habitat disturbances. Further studies must determine whether acid precipitation alters their habitat and food resources. The imitator salamander is not protected in North Carolina or Tennessee.

Taxonomy The imitator salamander was first described in 1927 and was named "imitator" because it resembles another salamander in the region, the red-cheeked salamander. No subspecies are recognized.

BLACK MOUNTAIN SALAMANDER *Desmognathus welteri*

BODY SHAPE
Slightly chubby;
broad snout

NUMBER OF COSTAL GROOVES
14

BODY COLOR
Brown

PATTERN
Irregular black to
dark brown spots or
streaks

BODY SIZE
Adults up to 6.75
inches in total
length

Description Black Mountain salamanders are brown with irregular black to dark brown spots or streaks on the back. The tail is flattened side to side, and its posterior half has a strong keel and is often lighter in color than the body. There is no sharp delineation between the color of the back and the sides. The belly is white with light brown to gray mottling. A light eye-to-jaw stripe is present, and the snout is broad and flattened. The tips of the toes are dark brown to black.

top An adult Black Mountain salamander

right Larvae and hatchlings have five to eight pairs of light spots on the back that fuse and become obscure with age.

Most Black Mountain salamanders are brown with irregular, sometimes faint, dark markings on the back. Their distribution is limited primarily to the Cumberland Mountains and Cumberland Plateau.

Larvae and Juveniles Larvae and hatchlings have five to eight pairs of light spots on the back between the front and hind limbs that fuse and become obscure with age. The belly is white with light mottling that intensifies with age.

Similar Species Black Mountain salamanders are difficult to distinguish from seal salamanders and dusky salamanders in the same area. Seal salamanders do not have dark toe tips and usually have a yellowish to whitish area underneath the tail, and the color of their back is usually distinguishable from the color on the sides. The belly of seal salamanders lacks mottling, and that of the dusky salamander has a reticulated or salt-and-pepper pattern.

Distribution Black Mountain salamanders are limited in distribution to the Cumberland Mountains and Cumberland Plateau of southwestern Virginia, southeastern Kentucky, north-central Tennessee, and southwestern West Virginia.

Habitat Adults and juveniles live in small to medium-sized streams with moderately steep gradients at elevations above 1,000 feet. They prefer streams with abundant rock cover that are shaded by forest canopy. Both age classes occupy wet stream margins and may wander into the adjacent forest during wet periods, although juveniles move farther than adults.

Behavior and Activity   These salamanders may be active in all but the coldest winter months; even then they may move around in water, but likely do not feed. They are not known to maintain territories but will move several yards up or downstream to find suitable habitat. They emerge to forage along stream margins and in the adjacent forest on wet nights in warm months.

Reproduction   Little is known about courtship and mating behaviors, but they are likely to resemble the mating behaviors of other salamanders in this group. Females with spermatophores have been found in May. Females lay 18–23 eggs in grapelike clusters in small cavities they construct in packed leaves, under rocks in stream banks, and in wet crevices in small cascades. Mothers remain

BLACK MOUNTAIN SALAMANDER   *Desmognathus welteri*

Black Mountain salamanders have broad, flat snouts and a light eye-to-jaw stripe.

until their eggs hatch in late summer. Larvae metamorphose into juveniles the following summer.

Food and Feeding Adults and juveniles eat adult and larval flies, beetles, and larval and winged wasps and bees, suggesting that they forage for insects out of water while perched on rocks or in the adjacent forest. Captive salamanders will eat earthworms.

Predators and Defense No predators have been reported, but spring salamanders, garter snakes, and watersnakes are probably among them. Antipredator behaviors probably include biting, fleeing to rock shelters, and body twisting and flipping.

Conservation The Black Mountain salamander is not protected by any state. It occurs in coal country, and so is subject to habitat loss and pollution from mining activities. Stream siltation and pollution from mine tailings are major threats. Mountaintop removal in which soil and rock are dumped downslope into streams destroys entire populations. Collection for fish bait by the "spring lizard" industry has been a serious threat but may be declining.

Taxonomy The Black Mountain salamander was first described to science in 1950.

BLACK-BELLIED SALAMANDERS

Black-bellied Salamander   *Desmognathus quadramaculatus*

Dwarf Black-bellied Salamander   *Desmognathus folkertsi*

Cherokee Black-bellied Salamander   *Desmognathus gunigeusgwotli*

Pisgah Black-bellied Salamander   *Desmognathus mavrokoilius*

Kanawha Black-bellied Salamander   *Desmognathus kanawha*

Nantahala Black-bellied Salamander   *Desmognathus amphileucus*

BODY SHAPE	NUMBER OF COSTAL GROOVES	PATTERN	BODY SIZE
Stocky; tail less than half as long as body	14	Brown with irregular rust-colored blotches	Adults up to 8 inches in total length
	BODY COLOR Olive, reddish brown, dark brown, or black; black belly		

top Black-bellied salamanders occasionally leave the water to wander in the forest.

opposite In some populations the body may be covered with golden to brassy flecks.

top Adults and most juveniles have black bellies.

bottom Black-bellied salamander larvae have small whitish gills but absorb them during metamorphosis 2–4 years after hatching.

Description The common name of these large, stocky salamanders comes from the black belly that is present in adults and most juveniles. Adults are black to dark brown with irregular yellowish green to rusty blotches. In some populations the body may be covered with golden to brassy flecks. Old individuals may be nearly entirely black both above and below. The sides have a double row of conspicuous light spots. The snout and feet are usually light brown, and the toe tips are black. The tail is sharply keeled, less than half the body length, and flattened side to side. Dwarf black-bellied salamanders are almost identical but are about 30 percent smaller (to 6 inches in total length), have a pattern of alternating dark blotches on the back, and have a black belly when the body length is 2 inches or less. Both species have a light eye-to-jaw stripe.

Larvae and Juveniles Larvae and juveniles of these species have a double row of light spots on the sides between the front and rear legs. Older larvae and juvenile black-bellied salamanders have a conspicuous reddish line down the top of the tail fin; the body of juveniles is often uniformly olive. Dwarf black-bellied salamanders lack the reddish line and are rarely greenish. Young juveniles have

- ● Kanawha black-bellied salamander *Desmognathus kanawha*
- ● Nantahala black-bellied salamander *Desmognathus amphileucus*
- ● Cherokee black-bellied salamander *Desmognathus gunigeusgwotli*
- ● Pisgah black-bellied salamander *Desmognathus mavrokoilius*
- ● dwarf black-bellied salamander *Desmognathus folkertsi*

Black-bellied salamander juveniles may possess gray mottling on their bodies and sides, but this pattern changes to patches and then fades out completely as they grow old.

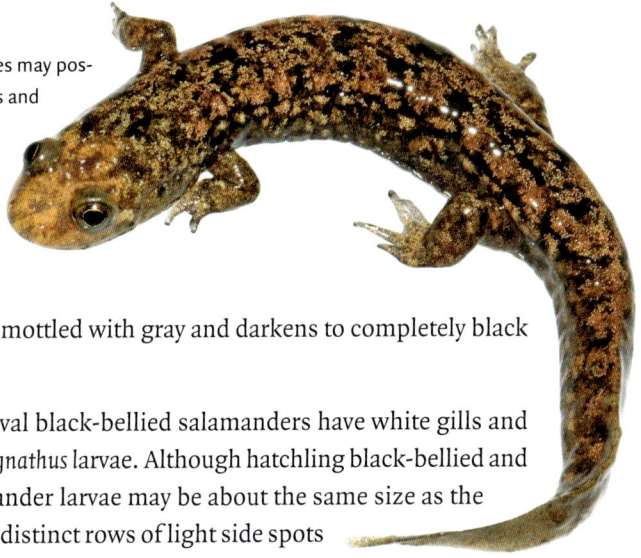

a white belly that becomes mottled with gray and darkens to completely black prior to sexual maturity.

Similar Species Older larval black-bellied salamanders have white gills and are larger than other *Desmognathus* larvae. Although hatchling black-bellied and dwarf black-bellied salamander larvae may be about the same size as the larvae of other species, the distinct rows of light side spots are distinctive. Other members of the genus usually do not have a black belly, although some old adults are black all over with a mottled belly and are difficult to distinguish. Shovel-nosed salamanders are mottled with much less black pigment, have a dusky belly that is not completely black, and the slope of the snout starts well behind the eyes.

Distribution Black-bellied salamanders occur from southern West Virginia through western Virginia and along the Appalachian and Blue Ridge mountains of North Carolina and Tennessee south to Georgia. Dwarf black-bellied salamanders are known from most of Georgia's Blue Ridge and adjacent areas in North and South Carolina.

Habitat These semiaquatic salamanders prefer high-gradient headwater and second-order mountain streams with abundant rock cover. They occasionally leave the water to wander in the forest. Larvae live in high-current riffles, particularly those with gravel-sand interstices. Juveniles are most abundant in flat, rock-strewn edge habitats with numerous small hiding places. Adults, particularly large black-bellied salamanders, often use burrows in stream banks.

Behavior and Activity Black-bellied salamanders are active year round if temperatures are warm enough to allow surface activity. These species are adept at wedging themselves between rocks and enlarging holes in dirt. They defend such burrows and crevices and use them as bases from which to make occasional forays.

Reproduction Courtship behavior has not been described for either species but is probably similar to that of other salamanders in this group, with the male enticing the female to pick up a spermatophore in her cloaca. Females lay as

DID YOU KNOW?

The primary prey of some salamanders is other salamanders. Black-bellied salamanders emerge from streams on rainy nights to forage on smaller species of salamanders along stream margins.

above Female black-bellied salamanders attach their eggs to the bottoms of rocks in shallow streams.

right Black-bellied salamanders commonly prey on other salamanders, affecting community structure and the presence of other species.

many as 80 eggs per clutch from April to July, attaching them to the underside of rocks in riffles or fast-moving streams as a single layer or in clumps. A female usually lays no more than one clutch every other year. The eggs hatch in summer and fall. The larvae of dwarf black-bellied salamanders take 2 years to reach metamorphosis; those of black-bellied salamanders take as long as 4 years.

Food and Feeding Adults and juveniles feed on a wide variety of invertebrates in water, but during wet periods they may leave their streams to forage on and in the leaf litter up to 30 feet away. Larvae eat aquatic prey such as mayflies, stoneflies, craneflies, and crayfish. Juveniles and adults feed mostly on terrestrial prey, including bees and wasps, butterflies and moths, beetles, centipedes, and spiders. Larger salamanders can take larger prey.

Predators and Defense Predators of black-bellied salamanders include shrews and garter snakes, but little is known about predator-prey relationships. Northern watersnakes, spring salamanders, aquatic invertebrates, and perhaps crayfish may be predators of larvae and small juveniles. Adults may on rare occasions bite a human finger but do not break the skin.

This dwarf black-bellied salamander from Union County, Georgia, has a pattern of dark blotches on its back.

Conservation The primary threats to black-bellied salamanders are alteration of natural streams resulting from acid precipitation and other atmospheric chemicals, and clear-cutting and lack of forest maintenance of watersheds above streams to prevent siltation and alteration of thermal qualities. Collection of these salamanders for fish bait (as "spring lizards") may once have caused population declines, but the practice has largely stopped and was probably never a widespread problem.

Taxonomy The black-bellied salamander was described in 1840. The dwarf black-bellied salamander was described as a new species in 2002. No subspecies are recognized for either, but recent molecular research indicates that D. *quadramaculatus* is a cryptic species comprising as many as six genetically distinct but morphologically indistinguishable populations. The authors of a 2022 paper have proposed dividing the black-bellied salamander into four new species based on nuclear and mitchondrial DNA evidence: the Nantahala black-bellied salamander (D. *amphileucus*) from the southern Appalachians in Georgia, South Carolina, and southern Tennessee and North Carolina; the Cherokee black-bellied salamander (D. *gvnigeusgwotli*) from the Great Smoky Mountains; the Pisgah black-bellied salamander (D. *mavrokoilius*) from North Carolina and Tennessee south of the Smokies and west of the French Broad River; and the Kanawha black-bellied salamanders (D. *kanawha*) from northern North Carolina, Virginia, and West Virginia. Time will tell if these proposed taxonomic changes will be accepted by the herpetological community. Scientists currently lack data on the ecological mechanisms, behaviors, and pheromones leading to the reproductive isolation of these populations, information that is needed to determine whether these newly described species are valid.

OCOEE SALAMANDERS

Ocoee Salamander   *Desmognathus ocoee*

Cherokee Mountain Dusky Salamander   *Desmognathus adatsihi*

Balsam Mountain Dusky Salamander   *Desmognathus balsameus*

Chattooga Dusky Salamander   *Desmognathus perlapsus*

BODY SHAPE	NUMBER OF COSTAL GROOVES	PATTERN	BODY SIZE
Slender and slightly flattened	14	Alternating light spots edged with black	Adults up to 4 inches in total length
BODY COLOR Brown to gray			

Description Ocoee salamanders have a light line between the eye and the back of the jaw. The base of the tail is slightly flattened (wider than tall), and the remainder is triangular. The pattern on the back is highly variable and usually includes a series of interconnected, yellow to orange, alternating, irregular spots with dark, wavy edging that extends well onto the tail. Some individuals have a straight-edged stripe. The body and tail are brown, gray, yellowish, or

top Body coloration is extremely variable among Ocoee salamanders from the same location and can change even more as individuals get older and larger.

Ocoee salamander larvae and hatchlings have reddish brown backs with light spots.

reddish and may be nearly black in old individuals. The belly is mottled with brown, gray, red, or yellow; old males may be nearly black. Some individuals have reddish cheeks or legs.

Larvae and Juveniles Ocoee salamander larvae (to 0.7 inches in total length) and hatchlings (to 1.1 inches in total length) have seven pairs of conspicuous alternating light spots on a reddish brown back. Juveniles are brownish to reddish with alternating light spots edged in brown to black.

Similar Species Spotted dusky salamanders have a keeled tail that is taller than wide at the base; the body is tan to brown with a wavy or smooth-edged yellowish to reddish brown stripe.

Distribution Ocoee salamanders are found from southeastern Tennessee and southwestern North Carolina south of the Pigeon River into northern South Carolina and Georgia. The Cherokee mountain dusky salamander (D. *adatsihi*) is found on the North Carolina–Tennessee border at the Smoky Mountains and Plott Balsam Mountains. The Balsam mountain dusky salamander (D. *balsameus*) is found in the Great Balsam Mountains of northern North Carolina. The Chattooga dusky salamander (D. *perlapsus*) is found from the tristate corner of

These small salamanders hide from predators in a wide variety of small places.

- Ocoee salamander
  Desmognathus ocoee

- Cherokee mountain
  dusky salamander
  Desmognathus adatsihi

- Chattooga dusky salamander
  Desmognathus perlapsus

- Balsam mountain
  dusky salamander
  Desmognathus balsameus

- *Desmognathus ocoee* complex

Females lay 12–28 eggs in July or August and stay with the eggs until they hatch in August or September.

Some Ocoee salamanders have red legs.

South Carolina, North Carolina, and Georgia; across the piedmont of Georgia to the Fall Line; and westward to adjacent Alabama.

Habitat Ocoee salamanders live in high-elevation seeps, first-order streams, wet rock faces, and mixed hardwood forests adjacent to streams. They avoid deep water and dry land, instead preferring spots where microhabitats interface. In hardwood forests they can often be found in wet mosses and under logs in moist leaf litter. Populations in low elevations (740 feet or less) are generally restricted to seepage areas and small streams; those in high elevations (to 4,600 feet) occupy the moist forest floor far from water.

Behavior and Activity Ocoee salamanders are active on land in all but the coldest months. They remain hidden during the day beneath cover or in rock face crevices and emerge at night to feed and search for mates. Adults actively defend their small territories against other adults. Adults and juveniles congregate deep in seeps and underground retreats in winter.

Reproduction Courtship and mating occur on land in spring and in late summer and fall. In July and early August, females lay 12–28 eggs in small clusters under rocks in seeps, in leaf litter, under moss, and in crevices in wet rock faces where shallow running water is present. Mothers remain with their eggs until they hatch in August to late September after a developmental period of 2–2.5 months. The larval period is 8–10 months; metamorphosis occurs in May and June when the larvae are about an inch long.

above The Ocoee salamander species complex exhibits some of the greatest color and pattern variation of all the eastern salamanders.

left Ocoee salamanders often climb vegetation at night during moist conditions to feed on arboreal invertebrates.

Food and Feeding Ocoee salamanders will eat virtually any small animal they can capture and swallow, including their own eggs and hatchlings. Prey of adult and juveniles includes flies, ants, beetles, wasps, spiders, mites, grasshoppers, and moths. Adults will climb and search for food in low, usually herbaceous, vegetation on wet nights.

Predators and Defense Known predators of Ocoee salamanders include ringneck snakes, garter snakes, watersnakes, black-bellied salamanders, seal salamanders, spring salamanders, small mammals, and birds that forage in leaf litter. When threatened they may autotomize their tail, flip their body, become immobile, or flee. Some of the more colorful individuals in this group may receive some protection by mimicking toxic newt efts.

Conservation These small salamanders are considered to be among the most abundant salamanders in the southern Appalachians and are not protected except where they occur on protected lands. Their fragile habitats are threatened, however, by timber removal and climate change that may alter hydrological patterns in the mountains. Continued monitoring of selected areas along elevational gradients may reveal population changes in response to regional climate changes.

Taxonomy The Ocoee salamander is part of the mountain dusky salamander complex and was first described in 1949. Numerous (as many as five) subgroups within this species occupy different mountain ranges. In 2022, the Ocoee salamander was divided into three additional species: the Cherokee mountain dusky salamander (*D. adatsihi*), the Balsam mountain dusky salamander (*D. balsameus*), and the Chattooga dusky salamander (*D. perlapsus*). As more research data is gathered additional species may be described in this complicated species complex.

Populations of this highly variable species complex occur on isolated mountain ranges. Modern taxonomic techniques have revealed that Ocoee salamanders, once considered a single species, are actually several separate species.

APALACHICOLA DUSKY SALAMANDER
Desmognathus apalachicolae

BODY SHAPE	NUMBER OF COSTAL GROOVES	BODY COLOR	BODY SIZE
Small; tail rounded and filamentous	14	Brown to nearly black	Adults up to 4 inches in total length

PATTERN
Series of reddish round spots

Description Apalachicola duskies are robust salamanders with rear legs that are larger than the front legs. A light stripe runs between the eye and jaw. The long tail is round at the base and tapers into a filament. Five to seven pairs of somewhat round cream, tan, red, or yellowish blotches run down the middle of the back and onto the tail on a black to brown background. The body spots may fuse to form a light stripe bordered on each side by black or dark brown. The top of the tail is the same color as the blotches. The head and sides of the body and tail are brown. Adult males, especially old ones, may be entirely black or brown. The belly is uniformly white, although some individuals have smudges of dark pigment. Females are more vividly patterned than males.

top The stout body, large hind legs, and light jaw stripe are typical of adult Apalachicola dusky salamanders. The tail is round at the base and tapers significantly at the tip.

Juveniles are lighter and have more distinct markings than adults.

Larvae and Juveniles The larvae have five to seven pairs of light blotches with tiny, white, sticklike external gills. Juveniles are colored and patterned as adults with yellow, brown, or red blotches but are brighter.

Similar Species The southern dusky is nearly uniformly black to dark brown with small white or reddish flecks concentrated on the sides and on the black belly. Spotted dusky salamanders have a brown back, with black markings in an irregular stripe edged in black, and a cream to tan belly with dark brown mottling.

Distribution Apalachicola dusky salamanders are found in southeastern Georgia from Upson County south to the center of the Florida panhandle.

Habitat These salamanders inhabit seeps and small headwater streams in the Chattahoochee and Ochlockonee river watersheds. They hide under leaves, logs, and rocks during the day and move and forage at night. They often sit in debris piles in very shallow water with only their head exposed in the air. The larvae remain in very shallow flowing water. Apalachicola duskies are rarely found away from water.

Apalachicola dusky salamanders hide under leaves, logs, and rocks during the day and move about at night to forage.

Behavior and Activity These salamanders are known to be primarily nocturnal, but their seasonal activity patterns are unknown. They may be active throughout the year, although they retreat underground in the hot summer months and move about on the surface only during rain. Adults are seldom found together, suggesting that they maintain territories.

Reproduction The timing of courtship and mating is not known. Females lay loose clusters of eggs under rocks, small logs, or other cover objects on stream

APALACHICOLA DUSKY SALAMANDER *Desmognathus apalachicolae*

Apalachicola dusky salamanders occur primarily in the moist environments of steep ravines associated with seepages and headwater streams from southeastern Georgia to the center of the Florida panhandle.

edges or in seeps in May and June. Clutch size has not been reported. Females remain with their eggs until they hatch in May through November, presumably to deter potential predators.

Food and Feeding Larvae and adults are entirely carnivorous and probably feed on small insects and other invertebrates.

Predators and Defense No predators have been documented, but they likely include several species of snakes, other salamanders such as red salamanders, snapping turtles, large frogs, shrews, raccoons, and other mammals. Apalachicola dusky salamanders rely on concealment and cryptic coloration to avoid predators. They can also leap by using their strong, enlarged rear legs.

Conservation The topography in the range of this small salamander affords some measure of habitat protection because the deep ravines are not suitable for houses or other buildings, although damming a stream to make a pond can destroy the habitat. The species has received no formal protection because it remains locally abundant.

Taxonomy The Apalachicola dusky salamander was not described until 1989. No subspecies have been described.

COMMON DUSKY SALAMANDERS

Spotted Dusky Salamander   *Desmognathus conanti*

Northern Dusky Salamander   *Desmognathus fuscus*

Santeetlah Dusky Salamander   *Desmognathus santeetlah*

Flat-headed Salamander   *Desmognathus planiceps*

Wolf Dusky Salamander   *Desmognathus lycos*

Piedmont Dusky Salamander   *Desmognathus bairdi*

Savannah Dusky Salamander   *Desmognathus campi*

Western Dusky Salamander   *Desmognathus catahoula*

Foothills Dusky Salamander   *Desmognathus anicetus*

Tilley's Dusky Salamander   *Desmognathus tilleyi*

BODY SHAPE
Robust; rear legs larger than front legs

NUMBER OF COSTAL GROOVES
14

BODY COLOR
Brown to gray

PATTERN
Highly variable, often with a broad reddish stripe on the back

BODY SIZE
Adults up to 5.5 inches in total length

top Spotted dusky salamanders occur from the Appalachian highlands to the Gulf of Mexico.

Description Northern dusky salamanders and the closely related species described here are medium-sized, stout-bodied salamanders with rear legs larger than the front legs. The base of the tail is taller than wide, and the remainder is triangular in cross section and keeled along the top edge. The eye-to-jaw stripe is prominent in all but old individuals. The base color is brown to gray. The back pattern, if present, is highly variable and darkens with age. Reported patterns include six or seven pairs of black, irregularly shaped spots; and a continuous dark yellowish to reddish brown stripe that may be straight-edged or wavy along each edge. The reddish stripe on the back usually extends onto the base of the tail. The belly is yellowish brown to cream with some black to gray mottling.

Spotted dusky salamanders have bolder markings on the back, and the spots may be red to golden. Santeetlah dusky salamanders have a poorly defined stripe on a brownish to greenish background; the underside of the tail may be yellowish. Flat-headed salamanders differ from northern duskies in tooth size

top Northern dusky salamander

bottom Santeetlah dusky salamander

top Spotted dusky salamander

bottom Flat-headed salamander

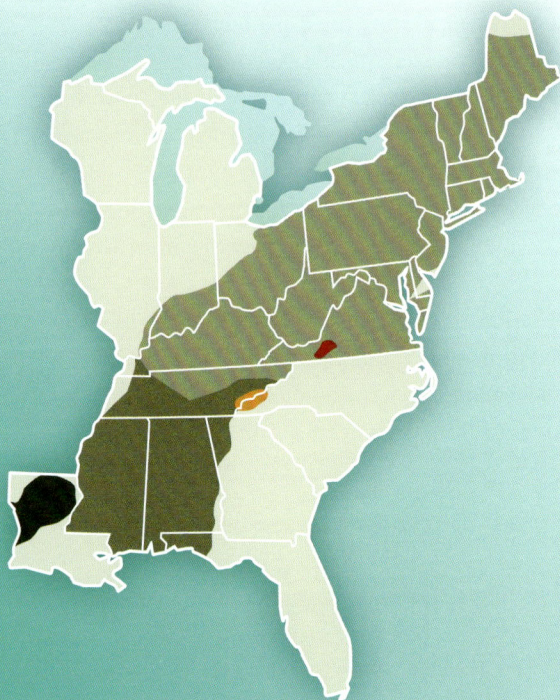

- northern dusky salamander
  Desmognathus fuscus
- western dusky salamander
  Desmognathus catahoula
- spotted dusky salamander
  Desmognathus conanti
- Santeetlah dusky salamander
  Desmognathus santeetlah
- flat-headed salamander
  Desmognathus planiceps

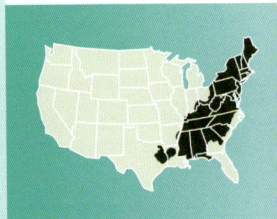

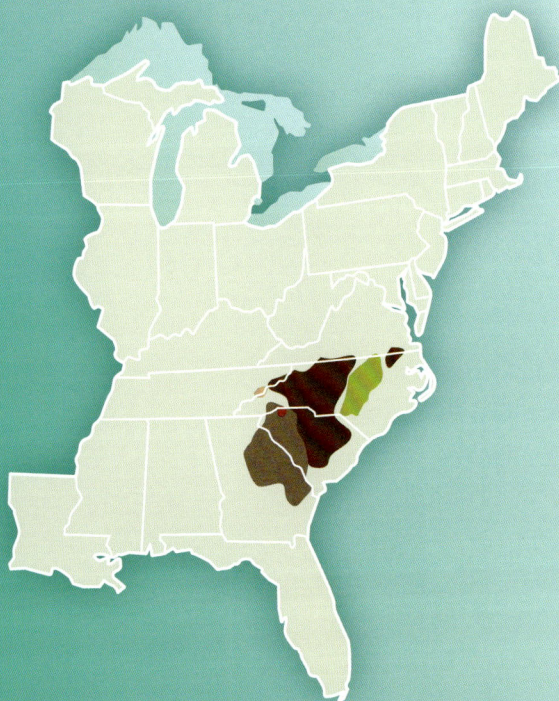

- Savannah dusky salamander
  Desmognathus campi
- wolf dusky salamander
  Desmognathus lycos
- Piedmont dusky salamander
  Desmognathus bairdi
- Tilley's dusky salamander
  Desmognathus tilleyi
- foothills dusky salamander
  Desmognathus anicetus

(D. planiceps has broader crowns and slightly larger teeth), but otherwise are nearly identical.

Larvae and Juveniles The larvae are small (less than an inch long) with five to eight pairs or sets of alternating light spots on a brownish background; external gills are present. Juveniles retain the spots until maturity, at which time they fuse into the broad stripe.

Common dusky salamander larvae have white gills that are absorbed during metamorphosis.

Similar Species Mountain dusky salamanders have a rounded tail. The dark pattern on the back of seal salamanders is reticulate (netlike) or a series of irregular black spots. Black-bellied salamanders have a black belly. Shovel-nosed salamanders have a paddle-like tail and two rows of light blotches on the back. Larval and juvenile desmogs are the most difficult to identify. Some species can be eliminated on the basis of locality. Consult the descriptions of each species in this book and study photographs in several sources to further narrow the possibilities.

Distribution Species of the northern dusky salamander group range from Nova Scotia south through Kentucky and Virginia into the Piedmont, the Blue Ridge and Appalachian mountains, and the Cumberland Plateau. They overlap and may hybridize with species of the spotted dusky salamander complex in South Carolina, northern Georgia, and Tennessee.

The Piedmont dusky and wolf dusky salamanders were recently delineated from the northern dusky salamander. The revised range of the northern dusky (D. fuscus) is from southern Canada through New England south to southern Tennessee, northwest North Carolina, Kentucky, Ohio and west to Indiana. The Piedmont dusky salamander is found in the Piedmont and adjacent Coastal Plain in North Carolina and across the upper Coastal Plain of South Carolina,

Northern dusky salamanders range north to Nova Scotia.

Santeetlah dusky salamanders occur in three mountain ranges in North Carolina, including the Great Smoky Mountains.

and the wolf dusky salamander is found from the Coastal Plain of southeastern Virginia and northeastern North Carolina through the Piedmont and Blue Ridge foothills of central North Carolina and a small portion of South Carolina.

Species of the spotted dusky salamander group occur from the Appalachian highlands west to the Mississippi River delta and south to the Gulf of Mexico on the Florida panhandle, Alabama, Mississippi, and Louisiana; isolated populations occur in the Georgia Coastal Plain, north-central Louisiana, and northeastern Arkansas. Four new species align themselves with the spotted dusky group. The Savannah or Camp's dusky salamander occurs in eastern Georgia and western South Carolina to the edge of adjacent North Carolina. The western dusky salamander is found in northern Louisiana and extreme eastern Texas. The foothills dusky salamander occurs in the foothills of north-central South Carolina and probably in adjacent North Carolina. Tilley's dusky salamander is found on the borders of North Carolina and Tennessee in and near the Great Smoky Mountains. With recognition of the new species, the range of D. conanti shrinks to southern Illinois, western Kentucky, Tennessee, and from the Tennesee–North Carolina border southward to northwest Georgia, Alabama, Mississippi, and Louisiana.

Santeetlah dusky salamanders are found in the Great Balsam Mountains of North Carolina and in the Great Smoky and Unicoi mountains of North Car-

Salamanders in this group, such as the spotted dusky salamander shown here, are active during all but the coldest winter months.

olina and Tennessee. Flat-headed salamanders are found in the Blue Ridge Escarpment east into the southern Piedmont of Virginia and possibly North Carolina.

Habitat Dusky salamanders prefer cool, well-oxygenated streams and their associated seepages. In the lowlands, this salamander appears to be restricted to seepage areas along small streams and the edges of swamps and slow-moving streams. Northern and Santeetlah dusky salamanders will disperse into the surrounding forest in rainy periods and can be found under logs and rocks.

Behavior and Activity Dusky salamanders are active during all but the coldest winter months, when they move into rocky crevices deep below the surface. From spring through fall they hide during the daytime under leaf litter, rocks, and logs in seeps and along stream margins, and forage and move at night. Communication is visual and by means of pheromones, but these salamanders do not mark territories. They remain under cover when relative humidity is low or it is not raining; the drier it is, the larger the rock they choose to hide under. High humidity and night rain bring out insects and other prey, and the salamanders come out to feed on them. Some individuals disperse up to 50–60 feet into the surrounding forest when the substrate is wet.

Reproduction Timing of courtship, egg laying, and other aspects of the life history of this wide-ranging salamander vary with elevation in the Southeast and with latitude in northern areas. Mating occurs in spring and fall. Some females appear to lay their eggs in the fall following mating; some do not lay eggs until the following May, June, or July. Each female deposits 11–36 eggs in grapelike clusters under rocks in water, in cavities under and inside logs, in leaf mats, or in clumps of moss within 6 feet of the stream edge. Santeetlah dusky salamanders lay about 3 more

Common dusky salamanders lay one to three dozen eggs in grapelike clusters in small depressions. The female remains with the eggs until they hatch.

eggs per female than northern dusky salamanders. Females in some areas lay eggs
every other year. Larvae hatch in about 2
months and remain in shallow water with
a gravel substrate to avoid the cannibalistic adults. The length of the larval period
depends on time of hatching, water temperature, and food availability. Larvae that
hatch in spring may metamorphose by the
following fall, but those that hatch in late
summer or fall overwinter as larvae and metamorphose more than a year later.
Size at metamorphosis is 1–2 inches.

Food and Feeding Larvae feed on copepods, fly and other insect larvae, collembolans, mites, and small fingernail clams. Juveniles and adults eat any animal
they can catch and swallow. The larger the salamander, the larger the prey item
it can consume. Aquatic and terrestrial insects and other invertebrates such as
larval and adult flies, caddisflies, true bugs, beetles, ants, spiders, caterpillars,
mites, and earthworms are common prey types.

Predators and Defense Known predators include northern watersnakes; common garter snakes; shrews; raccoons; skunks; bears; and black-bellied, red, and
spring salamanders. Cannibalism of larvae and adults occurs, but its frequency
is unknown. Dusky salamanders will bite potential predators but will first attempt to flee. The tail will break off if a predator bites it instead of the body.

Conservation Loss of headwater and seepage habitats as a result of agriculture,
urbanization, and road building has caused loss and decline of populations,
especially those outside protected areas such as state or national parks and
forests. Clear-cutting of hardwood trees affects water availability and thermal
characteristics in seeps and small streams. Collection of "spring lizards" for
fish bait can greatly reduce the size of some populations. Populations of these
long-lived (about a decade) salamanders that lay relatively few eggs cannot tolerate the removal of many adults. Mountaintop mining operations are always a
threat because they increase siltation and destroy habitat. Very little is known of
any aspect of the life history and population ecology of the species in this dusky
salamander group. Understanding the variation in timing of reproductive and
other critical events could provide insight into how to manage populations as
climate changes alter their habitats. Details of the ecology and life history of
the flat-headed salamander are almost completely unknown.

First described in 1955, the flat-headed salamander was elevated to full species status in 2008.

Taxonomy The northern dusky salamanders were described first in 1820. The spotted dusky was initially described as a subspecies of the northern dusky in 1958 but was elevated to a full species in 2002. The Santeetlah dusky was described in 1981. The flat-headed salamander was first described in 1955 but was not recognized as a full species until 2008.

The taxonomy of the common dusky salamanders has been long debated, and these salamanders have been acknowledged as a complex of multiple species that would eventually result in the recognition of several new species. The two major species groups (northern dusky and spotted dusky) were delineated in 2023 into a total of six new species based upon DNA analysis. The northern dusky group (D. fuscus) yielded two additional species: the wolf dusky salamander (D. lycos) and the Piedmont dusky salamander (D. bairdi). The other four new species align themselves with the spotted dusky group (D. conanti): the Savannah or Camp's dusky salamander (D. campi); the western dusky salamander (D. catahoula); the foothills dusky salamander (D. anicetus); and Tilley's dusky salamander (D. tilleyi).

SEAL SALAMANDERS

Seal Salamander   *Desmognathus monticola*

Talladega Seal Salamander   *Desmognathus cheaha*

BODY SHAPE	NUMBER OF COSTAL GROOVES	PATTERN	BODY SIZE
Robust; knife-edged tail tip	14	Irregular bold, dark reticulation or spots	Adults up to 5 inches in total length
	BODY COLOR Brown to gray		

Description Seal salamanders are stout-bodied, pop-eyed animals with a flattened tail that has a sharply defined keel on the last two-thirds of the upper surface. The brown to dark gray body has a bold pattern of black or brown markings that are wormlike, netlike, or in circles surrounding areas of background color. The belly is plain grayish cream without markings. A single row of small, irregularly shaped white spots may be present on each side between the front and hind legs. Older adults tend to lose the pattern on the back completely and turn purplish brown while their belly becomes gray or brown. Some populations may have dark brown or black toe tips.

top The black-and-brown body pattern, large pop eyes, and tail with a keel on the posterior portion are typical of seal salamanders.

top Larval seal salamanders have small white gills and 4 or 5 pairs of light spots on a light brown body.

bottom Juveniles are brown with 4 or 5 pairs of chestnut or orange-brown spots along the back.

Larvae and Juveniles Larvae have small white gills and four or five pairs of light spots on a light brown body. Hatchlings and juveniles are brown with four or five pairs of chestnut or orange-brown spots along the back. This pattern changes quickly into the adult form within a year.

Similar Species Northern dusky salamanders and other desmogs in that group have a salt-and-pepper belly and usually a reddish brown edge to the stripe on the top of the base of the tail. Black-bellied salamanders have a black belly as adults, but juveniles have yellow

flecking; both age classes have two rows of tiny white dots along each side. The back of shovel-nosed salamanders is mottled to resemble the pebble substrate in streams, and the sloping of the head starts behind the eyes; the head of seal salamanders starts sloping at the eyes. Black Mountain salamanders have a white belly with darker mottling.

below, left and right Seal salamanders exhibit a wide range of variation and may be dark or light in background color.

seal salamander
Desmognathus monticola

Talladega seal salamander
Desmognathus cheaha

Older individuals, like this Talladega seal salamander from Atlanta, Georgia, may be uniformly dark with little pattern.

Distribution The seal salamander is found throughout the Appalachian Mountains and adjacent foothills from Pennsylvania southward through West Virginia, Virginia, Kentucky, Tennessee, North Carolina, South Carolina, and into northern Georgia. The Talladega seal salamander is found from the Ridge and Valley and Upper Piedmont of northern Georgia across central Alabama to extreme northwest Florida, although the Florida population may longer exist.

Habitat Many cool freshwater streams with well-aerated water in hardwood forests in the Appalachian and Blue Ridge mountains and foothills harbor seal salamanders. Mountain streams with a rocky substrate and gravel bars are preferred over silted, slow-moving streams. Salamanders avoid lower reaches of streams that support fish, and juveniles remain in gravel areas to avoid the cannibalistic adults.

Behavior and Activity Adults emerge from the water at night and sit on the rocks they hide under in the daytime to wait for prey. Juveniles minimize predation by avoiding the large rocks used by adults. Seal salamanders are active year round in water, even when there is ice on the surface. During periods of heavy rains, some adults leave their stream and move a short distance into the adjacent forest, presumably to forage. During rain they occasionally climb 3–6 feet on tree trunks to forage. Males actively defend their shelter sites from other males.

Reproduction Mating occurs in spring and fall. Females produce 13–40 eggs, apparently annually, during mid-June to August and lay them as loose clumps or in a single layer attached to the underside of rocks in streams, in moss in water, in wet crevices, and in seepage banks. Eggs hatch in late summer to early fall, and the larvae overwinter in the stream. The external gills are absorbed 8–10 months later during metamorphosis.

Food and Feeding Adults sit atop their rock shelters at night and wait for invertebrates such as bugs, beetles, flies, ants, stoneflies, moths, beetles, crickets, mayflies, millipedes, and earthworms to come along. Juveniles actively move around to forage, but the adults usually stay in one place.

Predators and Defense Spring salamanders, several species of fish such as brook trout, watersnakes, and ringneck snakes eat seal salamanders. Their main defenses are their slippery skin and rapid escape by diving into the stream to hide under a rock.

Conservation Fishermen collect seal and other stream salamanders to use for bait (sometimes referred to as "spring lizards"), and overcollection in some

Seal salamanders are slippery and escape quickly by diving into streams and hiding under rocks.

areas may have contributed to the decline of this species. The collection of salamanders for fish bait is prohibited in Jefferson National Forest in Virginia. Acid precipitation has apparently caused their decline in some streams. Clearcutting around mountain streams alters the stream environment and contributes to population declines.

Taxonomy The seal salamander was described initially in 1916. No subspecies are recognized, but the seal salamander has been identified as a species complex comprising two or more species that look physically similar. In 2023 the seal salamander was delineated into two separate species: the seal salamander (*D. monticola*) and the Talladega seal salamander (*D. cheaha*).

MOUNTAIN DUSKY SALAMANDERS

Cumberland Dusky Salamander   *Desmognathus abditus*

Carolina Mountain Dusky Salamander   *Desmognathus carolinensis*

Allegheny Mountain Dusky Salamander   *Desmognathus ochrophaeus*

Blue Ridge Dusky Salamander   *Desmognathus orestes*

BODY SHAPE	**NUMBER OF COSTAL GROOVES**	**BODY COLOR**	**BODY SIZE**
Slender; rounded tail	14	Gray, brown, or yellow	Adults up to 4 inches in total length
		PATTERN	
		Light stripe bordered by black	

Description   These four closely related species look so much alike that knowing where the salamander came from is critical for accurate identification. The tail is round at the base, and the flattened posterior portion lacks a keel. A light line connects the eye and angle of the jaw. The pattern on the back ranges from a straight-edged, light-colored (light brown, gray, reddish, yellowish, or red) stripe to a series of interconnected, light-colored blotches with dark, wavy edg-

top Some Blue Ridge dusky salamanders have a straight-edged, reddish stripe down the back.

top The light line from the eye to the back of the jaw of this Carolina mountain dusky salamander is the signature trait of the dusky salamander group.   *middle* Although many of the dusky salamanders are similar in appearance, Cumberland dusky salamanders often have reddish spots on the back.   *bottom* This Allegheny Mountain dusky salamander has black chevrons or irregularly shaped black spots inside the light stripe down its back.

right The paired light spots of hatchling dusky salamanders change to stripes or blotches as they become juveniles.

below Juvenile mountain dusky salamanders like this one are found most often among damp leaves and along stream margins.

ing. The stripe usually extends onto the base of the tail. Narrow black chevrons or irregular black spots may be present inside the light stripe. The body and tail are brown to nearly black, and the belly is light in young salamanders and light gray to black in adults. Old males may be completely black. The straight and wavy-edged patterns occur in all four species; the straight-edged form predominates in the Allegheny mountain dusky. The eight pairs of reddish spots on the back of Cumberland dusky salamanders are usually in evidence even in older, dark adults.

Larvae and Juveniles Cumberland dusky salamander larvae and juveniles are unknown. Larval Allegheny and Carolina mountain dusky salamanders have a rounded snout, external gills, and five or six pairs of alternating black spots on the back. Juveniles usually have reddish or yellowish stripes or blotches on the back. Hatchling Blue Ridge dusky salamanders have not been described, but juveniles have brownish, reddish, or yellowish stripes.

Similar Species The tail of northern dusky salamanders, seal salamanders, and their close relatives is taller than wide at the base and has a keeled edge. Members of this group also tend to be more robust than the mountain dusky group. The tail of the closely related Ocoee salamander is wider than high at the base and appears somewhat flattened, and a series of alternating light spots bordered by black scalloping forms an irregular wavy or straight line down the back. Two-lined salamanders are yellow with two dark lines along the back, and they lack the light eye-to-jaw stripe, as do all of the woodland salamanders in the genus Plethodon.

Distribution The Cumberland dusky is known only from seven counties in the Cumberland Plateau in central Tennessee. Carolina mountain dusky salaman-

MOUNTAIN DUSKY SALAMANDERS

- Cumberland dusky salamander
  Desmognathus abditus
- Carolina mountain dusky salamander
  Desmognathus carolinensis
- Allegheny mountain dusky salamander
  Desmognathus ochrophaeus
- Blue Ridge dusky salamander
  Desmognathus orestes

Mountain dusky salamanders typically remain active on or near the surface from March through October.

ders are found in the Blue Ridge, Black, Bald, and Unaka Mountains in western North Carolina and eastern Tennessee. Allegheny mountain dusky salamanders occur in the highland regions of the Adirondack Mountains in Quebec, Canada, southward to southwest Virginia and Tennessee. The Allegheny mountain dusky salamander hybridizes with the Cumberland dusky salamander in Morgan County, Tennessee. Blue Ridge dusky salamanders are limited to the Blue Ridge Mountains in southwestern Virginia and northwestern North Carolina. An apparently isolated population occurs in northeastern Alabama.

Habitat   Mountain dusky salamanders occupy cool, high-elevation seepage areas, first-order streams, wet rock faces, and mixed hardwood forests adjacent to streams. They avoid deep water and prefer the interface between flowing water and dry ground such as that found in wet mosses and under logs in moist leaf litter in forests. Allegheny mountain dusky salamanders are often highly terrestrial and can be found great distances from water. Juvenile mountain dusky salamanders are found most often in damp leaves and along stream margins.

Behavior and Activity   Mountain dusky salamanders are active on or near the surface in all but the coldest months—that is, generally from March through October. They remain active in seepage areas and springs even in winter. Their movement from streams and seeps into the surrounding forests coincides with rainfall; in dry conditions they limit their activity to wet habitats. Low-elevation populations remain close to seepages and wet rock faces, but those in high elevations wander into the surrounding forest.

Reproduction   Mating occurs in water or in moist spots on land from September through June in all but the coldest months. Females lay 5–37 eggs in grapelike clusters under rocks and logs; in moss, mud, and seepage banks; and in crevices on wet cliffs with flowing water. The timing of hatching varies with elevation, latitude, and food availability. Hatchlings have been found in fall and spring. The larval period likewise can be short (1–3 weeks) or long (8–10 months). Metamorphosis, consisting primarily of loss of gills, occurs about 7–10 months later, typically around May and June, again depending on location.

Food and Feeding The timing of metamorphosis coincides with the abundance of small invertebrates in shallow water. Juveniles and adults are generalist predators that consume a wide variety of prey, including ants, beetles, grasshoppers, caddisflies, spiders, mites, moths, and worms. Adults occasionally eat eggs, hatchlings, and juveniles of their own species.

Predators and Defense Known predators include ringneck snakes, garter snakes, watersnakes, spring salamanders, small mammals such as shrews, and turkeys and other birds that forage in leaf litter. These salamanders may bite, break off the tail, become immobile, or flee when threatened. Some of the more colorful individuals in this group may receive some degree of protection from mimicking toxic newt efts.

Conservation A large portion of the ranges of these four species in the Southeast occurs on public lands such as state and national parks, forests, and wildlife refuges. However, clearing land for timber threatens the environmental integrity of the headwater streams and seeps these salamanders require. Subsequent siltation and removal of the protective canopy can result in higher temperatures. Mountaintop removal eliminates entire populations. Climate change that increases temperatures in high elevations will also threaten seepage areas and other microhabitats. None of these salamanders are protected legally except by state regulations that limit the number that can be collected. The small range of the Cumberland dusky salamander makes that species vulnerable to habitat loss or alteration, and its conservation status needs immediate attention. Local climate change is expected to cause mountain species such as these to migrate upslope and reduce their ranges, and local populations may be extirpated.

Taxonomy Until 1996 these four species were considered one species, the Allegheny mountain dusky salamander, described in 1859. The Carolina mountain dusky salamander was described in 1916 as a subspecies, the Blue Ridge dusky salamander in 1996, and the Cumberland dusky salamander in 2003. No subspecies are currently recognized. The species in this complex can be distinguished from each other only by geographic location and genetic analysis.

BROWNBACK SALAMANDER *Eurycea aquatica*

BODY SHAPE
Small and stout with
short legs and tail

**NUMBER
OF COSTAL
GROOVES**
13–15

BODY COLOR
Medium brown to
coppery brown to
dark brown

PATTERN
Thin brown stripe on
the tail with the tip
yellowish brown

BODY SIZE
Adults up to 3.7
inches in total
length

Description The brownback salamander is a small, brownish salamander with a tail less than half its total length. The back is brown to coppery to dark brown with a stripe pattern but no distinct lateral stripe. Some individuals are yellowish, similar to two-lined salamanders. The belly is pale yellow and lacks a pattern. Males are the same size as females but usually have wider, more robust heads. They typically do not possess elongated cirri.

Larvae and Juveniles Larvae are dusky brown and reach a maximum length of 1.18 inches from snout to vent, with a streamlined body and short, bushy gills. They are very similar to the larvae of southern and Blue Ridge two-lined salamanders. At least one population in Bradley County, Tennessee, appears

top Brownback salamanders are small, semiaquatic brook salamanders.

A brownback salamander larva in its second year

to contain neotenic individuals with the larvae being larger than those at other population sites. Larvae are the most commonly encountered life stage of the species.

Similar Species The brownback salamander is similar in appearance to the more common southern two-lined salamander and Blue Ridge two-lined salamander, as well as the less common and more geographically restricted Ju-

BROWNBACK SALAMANDER *Eurycea aquatica*

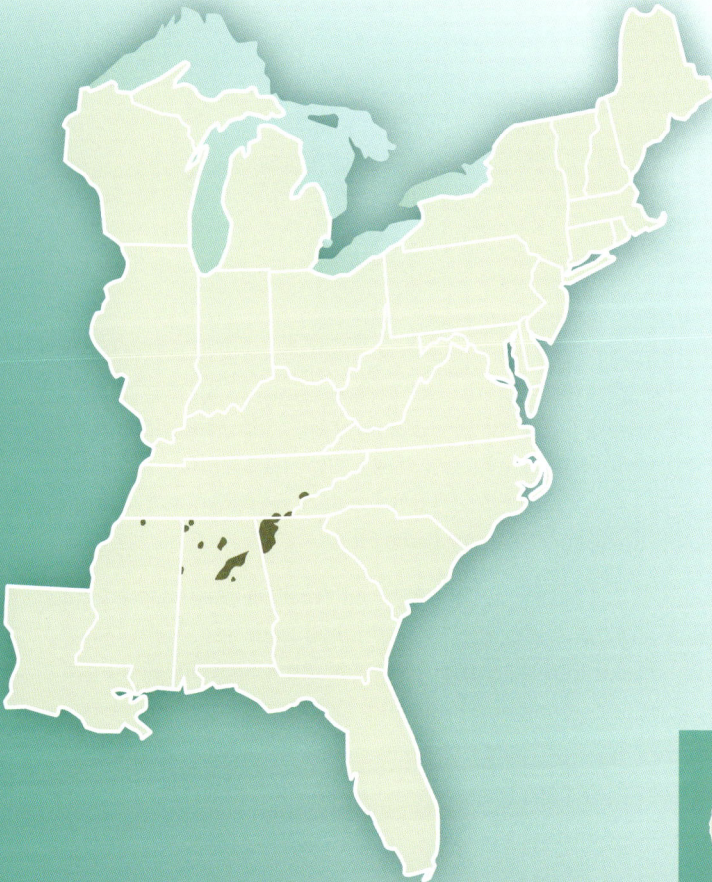

naluska salamander. The brownback salamander is more robust than these other species, possessing a wider head, reduced trunk, and shorter tail. It is also usually darker in color, with a brown back and darker brown to black sides. In Georgia, this species frequently co-occurs with the southern two-lined salamander, even under the same rock.

Distribution The brownback salamander's range is disjunct across north-central Alabama, northwestern Georgia, and several isolated localities in southeastern Tennessee (Bradley and Sevier Counties). Surveys are underway to see if the species has a more extensive distribution in the Cumberland Plateau and Valley and Ridge physiographic provinces in eastern Tennessee, Virginia, and Kentucky.

Habitat Most known sites are near or around springs that often have floating vegetation such as watercress (*Nasturtium officinale*). Further studies have found brownback salamanders to occur in small, rocky streams not associated with springs. Larvae are common in the known springs, and it is not known if the adults disperse to more terrestrial habitats or remain at the springs.

Behavior and Activity Little is known of the natural history of brownback salamanders. Some evidence indicates that adults spend nonbreeding seasons in nearby forests and migrate into the springs and/or streams to breed.

Reproduction Breeding males and gravid females are found from January to March. Males are often found with bite scars or nipped tails, indicating male fighting for breeding rights. Both males and females can be found under surface cover (logs, rocks, and sometimes artificial objects) in early March and April attending egg clutches of 79–100 eggs. Observations suggest that some

A gravid female brownback salamander from Bartow County, Georgia

A gravid female (top) and male (bottom) brownback salamander

nests are guarded only by females and some only by males whereas others are unattended. Eggs hatch in April and since two different larval size classes have been found together it is deduced that the larval stage is two years.

Food and Feeding These salamanders probably have a diet similar to the southern two-lined salamander. Food items probably include earthworms, isopods, millipedes, beetles, flies, midges, spiders, and aquatic insects such as caddisfly larvae and stonefly nymphs. Larvae probably feed on tiny aquatic insects, small worms, and tiny crustaceans.

Predators and Defense Specific predators have not been reported but likely include diving beetles, small snakes, turtles, and shrews. No antipredator defenses have been documented.

Conservation Little is known about the population status of the brownback salamander. Since it lives in small, isolated springs over much of its known range, it is likely vulnerable to agricultural and urban development, runoff, pollution, siltation, and the depletion of groundwater because of its removal from aquifers for agricultural purposes.

Taxonomy Although originally described in 1963, this species was long considered a southern two-lined salamander, a status that was questioned in 2009 as a consequence of molecular evidence. Its full species status was confirmed in 2014.

CAROLINA SANDHILLS
SALAMANDER *Eurycea arenicola*

BODY SHAPE	NUMBER OF COSTAL GROOVES	BODY COLOR	BODY SIZE
Small and thin; slender, laterally compressed tail	14	Orange or reddish orange	Adults up to 3.5 inches in total length

PATTERN
Small, diffuse black spots but no stripes

Description The Carolina Sandhills salamander is a small *Eurycea* with both the back and belly usually orange, reddish, or orange red. Small black spots on its back are never organized into a mid-dorsal stripe. Dorsolateral stripes are either indistinct or lacking. Sexually active males possess cirri, mental glands, and enlarged teeth in the front of the upper jaw.

Larvae and Juveniles Larvae have cinnamon brown dorsal surfaces, including on the tail and limbs. They have scattered brown spots and flecking on both the head and body. They have nine pairs of dorsolateral brown spots, each with a small but distinct cream spot, that extend from the rear of the head to the base of the tail. Gills have red gill stalks and the main gill stem is pink. Larvae are

top The Carolina sandhills salamander was described in 2020.

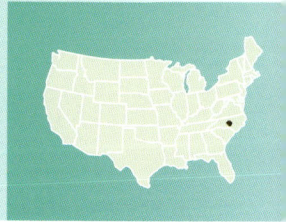

The Carolina sandhills salamander is a member of the two-lined salamander species complex inhabiting slow-moving blackwater streams.

commonly found in leaf beds, aquatic vegetation, and root beds in seeps and streams and have been collected every month of the year, indicating that the larval period before metamorphosis is at least one year. Juveniles are brown to olive brown with dorsal and lateral surfaces resembling the adult pattern.

Unlike the northern or southern two-lined salamanders, the Carolina Sandhills salamander is usually reddish or orange rather than yellow in overall coloration.

Similar Species The Carolina Sandhills salamander is smaller but similar in appearance to several species of the two-lined salamander species complex, but its geographic range only overlaps that of the southern two-lined salamander. It can be distinguished by its lack of distinct, continuous dorsolateral stripes and its lack of prominent dorsal spots that form a mid-dorsal stripe. Their bodies are reddish or orange rather than yellow in overall coloration. Carolina Sandhills salamander males are on average smaller than southern two-lined salamander males. The dwarf salamanders occur within the range of the Carolina Sandhills salamander but have only four toes per foot.

Distribution The Carolina Sandhills salamander has a geographic range limited to the Sandhills physiographic region of North Carolina and is documented from Harnett, Hoke, Montgomery, Moore, Richmond, Robeson, and Scotland Counties. It likely occurs in Cumberland County, but this has yet to be verified.

Habitat The Carolina Sandhills salamander inhabits slow-moving blackwater streams, often with sandy substrate, that flow through hardwood corridors of the Sandhills–Longleaf Pine ecosystem. It is associated with small streams and springs with sand, gravel, clay, and detritus at elevations of 167–492 feet.

Behavior and Activity Adults have been found at the surface during every month of the year, either crossing roads on rainy nights or positioned under surface debris such as logs, leaves, and/or pine needles. They appear to migrate to their breeding sites from December to February.

Reproduction Little is known about the reproductive ecology of this species, but gravid females have been found in late fall and early winter. Eggs are deposited during the winter and hatch into aquatic larvae in late winter to early spring. Natural nests have been found in submerged root masses at the edges of streams in February and March. Limited observations suggest females stay with their nests until hatching.

Food and Feeding Very little is known about the diet of this recently described species, but it is assumed to be similar to other small *Eurycea*. Carolina Sandhills salamanders have been found foraging at the surface, and their diet probably consists of a variety of small insects, ticks, spiders, mites, earthworms, isopods, springtails, flies, and ants.

Predators and Defense No predation has been observed, but shrews, ground foraging birds such as turkeys, larger salamanders, ringneck snakes, and garter snakes are probable predators. Other *Eurycea* avoid predation by staying still or fleeing, and most can break off their tail to escape.

Conservation The Carolina Sandhills salamander appears to be rather abundant in suitable habitat but has a very small geographic range (endemic to the Sandhills of North Carolina). It appears under the designation w3 (species that are poorly known) on a watch list maintained by the North Carolina Natural Heritage Program.

Taxonomy *E. arenicola* was described in 2020 based on ecological and morphological data as well as mitochondrial and nucleic DNA evidence. The member of the two-lined salamander species complex closest in geographic range to the Carolina Sandhills salamander is the southern two-lined salamander, yet this species is not the closest relative to the Carolina Sandhills salamander based on DNA evidence.

TWO-LINED SALAMANDERS

Northern Two-lined Salamander   *Eurycea bislineata*

Southern Two-lined Salamander   *Eurycea cirrigera*

BODY SHAPE
Small and slender

NUMBER OF COSTAL GROOVES
13–16

BODY COLOR
Yellow or light brown

PATTERN
Dark stripe on either side of the back

BODY SIZE
Metamorphs 1.6 inches; adults up to 4.7 inches in total length

Description Adults of both species are usually less than 4 inches in total length but can be almost 5 inches long. They are yellow, sometimes brownish yellow, with black or dark brown lines that run down the sides and onto the tail. Tiny black spots may be visible on the back. Individuals in some populations are uniformly brown. The belly is translucent and plain yellow. During the breeding season males commonly have a pair of cirri extending downward from the outer edge of the upper lip below the nostril. Northern and southern two-lined salamanders can be differentiated by the number of costal grooves (the north-

top The colorful southern two-lined salamander is found in all of the southeastern states as well as parts of Illinois, Indiana, and Ohio.

left The tiny aquatic larvae of two-lined salamanders have light spots along the sides instead of lines.

below Northern two-lined salamanders are often brownish yellow with a dark line down each side and tiny black spots on the back.

ern species typically has 15 or 16 and the southern typically has 13 or 14) and by location; their ranges overlap in the Southeast only in Virginia.

Larvae and Juveniles The aquatic larvae are tiny at hatching (less than 0.5 inches) and have obvious red gills. Larvae are dull gray or brownish with light spots along the sides and back and a clear belly. The end of the head is blunt. A clear tail fin is visible from above the hind legs to the end of the tail. Juveniles resemble adults.

Similar Species The Blue Ridge two-lined salamander is often more orange and the dark stripes become broken about halfway down the tail, but they remain complete for most of the tail length in the northern and southern two-lined salamanders; geographic distribution is an effective means of differentiating most specimens because of the limited areas of overlap. Three-lined salamanders have a dark line down the middle of the back, and the dark side stripes of long-tailed salamanders are more broken and irregular than those of two-lined salamanders. Adult three-lined and long-tailed salamanders are larger than either of the two-lined species. Most dusky salamanders that share habitats with two-lined salamanders have a brown or black body without distinct dark stripes and have a light mark behind the eye.

northern two-lined
salamander
Eurycea bislineata

southern two-lined
salamander
Eurycea cirrigera

Distribution Two-lined salamanders are widespread east of the Mississippi River. The northern two-lined salamander occurs from southeastern Canada through all of New England and south to West Virginia and northern Virginia. The ranges of the southern and northern species merge in Virginia in a narrow zone running from southern Rockingham County in the Shenandoah Valley northeastward to southern Prince William County in the Coastal Plain. The southern two lined salamander occurs from Illinois, Indiana, and southern Ohio southward to southern Georgia and the Florida panhandle.

A pair of courting southern two-lined salamanders along an urban stream in Atlanta, Georgia

Habitat Two-lined salamanders are associated with small streams or seepage areas bounded by forest habitat. Adults are most frequently seen in late winter and early spring in wooded stream habitats, often far from the stream. Both species occur in urban and suburban streams if rocks and other hiding places are abundant.

Behavior and Activity Some adults may be present under streamside rocks and logs at any time of year, but they are especially prevalent in March and April during the breeding season. They are most active in early evening. Juveniles and adults may live in drier forested areas far from the water, although they usually remain under moist ground cover or in burrows. Adults have been reported to be territorial. Some two-lined salamander populations spend most of the year in wooded habitats and migrate to stream pools for breeding.

Reproduction Two-lined salamander reproductive activity has been observed in most months of the year and varies according to geographic location, habitat type, and seasonal rainfall and temperature. Males in some populations develop cirri prior to breeding. Whether courtship and mating are primarily terrestrial or aquatic is unknown for most populations. A female will lay as few as 15 to more than 100 eggs, which she attaches beneath rocks or

A female two-lined salamander may lay up to 100 eggs, attaching them beneath rocks or logs in slow-moving water. She remains with her eggs until they hatch.

logs in slow-moving water. She remains with the eggs until they hatch, which may be about a month to more than 2 months. The larvae remain in the water for more than a year until they metamorphose.

Food and Feeding These small salamanders eat a wide variety of terrestrial invertebrates including earthworms, sowbugs, millipedes, beetles, snails, flies and midges, wood roaches, spiders, and ticks; and aquatic prey such as caddisfly larvae, stonefly nymphs, and fish eggs.

Predators and Defense Numerous predators have been documented, including ringneck snakes and garter snakes, screech-owls, black-bellied salamanders, and trout. Defensive maneuvers include remaining immobile, running, and tail waving. The tail will break off if a predator grabs it, allowing the salamander to escape. Secretions from skin glands make two-lined salamanders distasteful to some predators, such as certain species of shrews.

Conservation Two-lined salamanders are geographically widespread and are usually present or even abundant in woodland stream habitats in urbanized areas. Local populations are susceptible to activities that result in excessive stream pollution, heavy siltation, or degradation of the terrestrial landscape.

Taxonomy The northern two-lined salamander was described in 1818, and the southern two-lined salamander in 1830. The southern two-lined salamander was considered a subspecies of the northern two-lined salamander until 1987, as was the Blue Ridge two-lined salamander (E. wilderae). The two-lined salamander is apparently a variable complex. It may be divided into additional species once more research is conducted to reveal genetic affiliations.

DID YOU KNOW?

Senescence is not known to occur among salamanders. Some species are known to live more than a decade in the wild and more than a quarter of a century in captivity.

BLUE RIDGE TWO-LINED SALAMANDER   *Eurycea wilderae*

BODY SHAPE
Small and slender

NUMBER OF COSTAL GROOVES
14–16

BODY COLOR
Bright yellow to orange

PATTERN
Dark stripes on sides that break into blotches about halfway down the tail

BODY SIZE
Metamorphs 1.1 inches; adults up to 4.75 inches in total length

Description Blue Ridge two-lined salamanders are bright yellow, yellowish orange, or orange. Broad, well-defined dark lines run down the sides and onto the tail, breaking up midway down the tail into blotches. The back has black spotting. The belly is plain yellow. The total length is commonly 3.5 to more than 4 inches and may reach 4.75 inches. Males often have cirri and enlarged head muscles during the breeding season.

Larvae and Juveniles The aquatic larvae have gills and a noticeable clear keel on the tail; they are about 0.4 inches long at hatching and slightly more than an inch long at metamorphosis. The larvae can be dull brown, gray, or greenish yellow with light spots on the back and sides. They metamorphose after 1

top Blue Ridge two-lined salamanders have black spots on the back and well-defined dark lines that run down the sides and onto the tail.

year in warmer streams but may take up to 2 years in colder ones. The juvenile stage lasts about a year. Juveniles resemble adults.

Similar Species The Blue Ridge two-lined salamander can be distinguished from the southern two-lined salamander in areas where the two might overlap geographically by the latter's less well-defined side stripes, which usually remain unbroken on the tail. Other salamanders within the species' geographic range include the much larger three-lined salamander, which has a dark center line down the back; and the long-tailed salamander, whose side stripes are not as well defined. Other salamanders

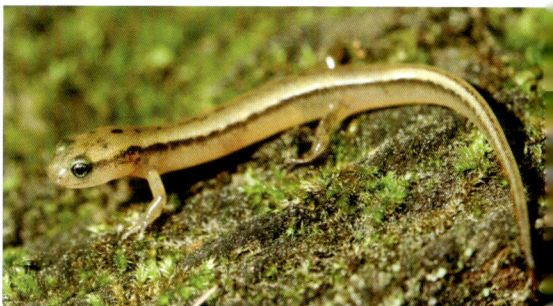

Blue Ridge two-lined salamander larva (*top*) and metamorph (*bottom*)

found in the habitats where Blue Ridge two-lined salamanders live do not have the bright yellow or orangish body color along with distinct side stripes. Dusky salamanders are mostly brown or black and have a light mark behind the eye.

Distribution Blue Ridge two-lined salamanders occur in the southern Appalachians in the Southern Blue Ridge Mountain physiographic province from White Top Mountain, Virginia, across North Carolina, and into about a dozen counties in northeastern Georgia, including areas within the Upper Piedmont province. The range is completely surrounded by that of the southern two-lined salamander.

Habitat These salamanders occupy mountain stream systems in mixed hardwood forests and spruce forests to elevations above 6,000 feet. They often occur in the terrestrial habitat, sometimes several hundred feet from streams.

Behavior and Activity During nonbreeding seasons Blue Ridge two-lined salamanders sometimes migrate far into forested areas and remain under ground litter such as leaves and dead bark. They can be found beneath rocks along mountain streams at most

Blue Ridge two-lined salamanders often occur in hardwood forests and spruce forests, sometimes several hundred feet from streams.

Females of this species remain
with their eggs until they hatch.

times of the year but are less abundant during
summer.

Reproduction The timing of reproduction varies according to elevation, stream temperature,
and other factors not yet fully determined. As
early as October adults begin migrating from
woodland habitats to streams where they congregate for courtship and mating. Some males
develop prominent cirri during the breeding pe-

Defensive strategies of Blue Ridge two-lined salamanders include waving the tail, which can break off.

riod; the head becomes enlarged in others. Each female lays a clutch of 8–56 eggs in late winter to early spring, gluing them to the underside of rocks in the water. Several females and egg clusters can sometimes be found under the same rock. The female remains with her eggs until they hatch in late spring to early summer, the timing depending on water temperature and time of egg laying.

Food and Feeding Blue Ridge two-lined salamanders probably eat a wide variety of small terrestrial and aquatic invertebrate prey such as earthworms, sowbugs, flies, spiders, and the larvae of aquatic insects.

Predators and Defense Larger salamanders, including spring salamanders and black-bellied salamanders, and garter snakes are known predators. Red salamanders, shovel-nosed salamanders, and ringneck snakes probably eat them, also. Defensive strategies are variable and probably include remaining immobile when exposed; rapid retreat under cover objects; and waving the tail, which can break off.

Conservation Blue Ridge two-lined salamanders have historically been among the most common stream salamanders within their geographic range and continue to thrive in most areas that have not suffered undue aquatic pollution or siltation. Thermal changes to environments at high elevations that could alter stream temperatures might affect populations of these salamanders.

Taxonomy The Blue Ridge two-lined salamander was first described in 1920 as a subspecies of the northern two-lined salamander and was elevated to full species status in 1987. The entire two-lined salamander species complex, including the Blue Ridge two-lined salamander and southern two-lined salamander, will probably yield several new species as mitochondrial and nucleic DNA research continues.

THREE-LINED SALAMANDER *Eurycea guttolineata*

BODY SHAPE
Long and slender;
broad head

NUMBER OF COSTAL GROOVES
13 or 14

BODY COLOR
Dull yellow to tan

PATTERN
Dark stripe on each
side and down cen-
ter of body

BODY SIZE
Metamorphs 1.2
inches; adults up to
7.9 inches in total
length

Description The body is usually yellow but can be tan, brownish, or orangish.
A broad, distinct dark brown or black line runs the length of the body and tail
on each side. The line may become infused with lighter color on the tail but is
still apparent. A dark line, which may be broken or well defined and continuous, extends down the center of the upper body from behind the head to the
base of the tail. The belly is mottled with gray pigment on a cream to yellow
background. Adults are often more than 6 inches in total length and can be
nearly 8 inches long. The tail is long, sometimes making up almost two-thirds
of the total body length.

Larvae and Juveniles The gilled aquatic larvae are about 0.5 inches long at
hatching and are grayish to cream-colored with stripes and a clear tail fin. The

top Three-lined salamanders have long tails, sometimes makng up almost two-thirds of
their body length.

Three-lined salamanders are recognizable by the broad, distinct dark line running the length of the body and tail on each side and the dark line down the center of the back.

larvae metamorphose at a length of 0.9–1.3 inches after 4–6 months in warmer areas, or after more than a year in colder ones. Juveniles resemble adults in pattern and color.

Similar Species The three dark longitudinal body stripes on a yellowish body distinguish the three-lined salamander from other species within the geographic range. Some long-tailed salamanders have a central row of spots down the back and can resemble three-lined salamanders, but the two species do not overlap in most parts of their respective ranges.

Distribution The geographic range of the three-lined salamander includes the portion of Virginia east of the Blue Ridge Mountains and most of the Carolinas and Georgia to the Florida panhandle, most of Alabama and Mississippi to the upper southeastern corner of Louisiana, western Tennessee, and southwestern Kentucky. The species is noticeably absent from most of the Coastal Plain of North Carolina, the southeastern and northwestern corners of Georgia, and all of peninsular Florida.

Three-lined salamanders' bellies are mottled gray on a cream to yellow background.

Habitat Three-lined salamanders are found in seepage areas, stream edges, and in river floodplains and swamps. They are

Three-lined salamanders are wide-spread geographically, being found in parts of every southeastern state, and are abundant in many areas.

Three-lined salamanders are active above ground at night, especially during rains, and remain beneath cover during the day.

sometimes found in the crevices of rock outcrops some distance from water. They are primarily terrestrial but do not venture far from the aquatic sites where they lay their eggs, and they remain in moist areas beneath ground litter of leaves, logs, and dead bark, or under rocks.

Behavior and Activity These salamanders are active above ground at night, especially during rain, and remain beneath cover during the day. Individuals are most often seen during warmer months but can be found under ground litter throughout the year. Unlike many other terrestrial salamanders that live in wooded areas and breed in aquatic habitats, three-lined salamanders usually stay in the vicinity of the water. They are not overtly territorial as many other stream salamanders are.

Reproduction Surprisingly little is known about the courtship and mating patterns of three-lined salamanders, which are relatively abundant over a wide geographic range. Eggs are laid in the late fall or winter, mainly in December, but apparently not in the large identifiable clusters characteristic of other stream salamanders. Females are believed to lay about a dozen eggs underwater but do not stay with them until hatching.

Food and Feeding The diet consists of a wide variety of small terrestrial invertebrates, including several types of insects such as ants, flies, beetles, grasshoppers, and others; spiders; snails; worms; and millipedes.

Predators and Defense Direct predation has not been observed, but ringneck snakes, garter snakes, and watersnakes would readily eat these salamanders, as would larger salamanders that live in the same habitat. If they are unable to escape by running or swimming away, three-lined salamanders sometimes assume a defensive behavior by hiding the head beneath the tail, which they curl and slowly wave.

Conservation Three-lined salamanders are widespread geographically and abundant in many areas. Local populations may be susceptible to habitat alterations such as the clearing of bottomland hardwood forests, siltation, and pollution.

Taxonomy The three-lined salamander was described in 1838 and was considered a subspecies of the long-tailed salamander (*Eurycea longicauda*) until 1998, when genetic research indicated that three-lined and long-tailed salamanders do not normally hybridize in regions of overlap.

LONG-TAILED SALAMANDER   *Eurycea longicauda*

BODY SHAPE	NUMBER OF COSTAL GROOVES	PATTERN	BODY SIZE
Long and slender; broad head	13 or 14	Black spots on body that may be in rows but do not form distinct lines	Metamorphs 1 inch; adults up to 7.75 inches in total length
	BODY COLOR Orangish yellow or brownish yellow		

Description   The basic body color is yellow, often with a tinge of orange or brown. The belly is yellowish without a pattern. Black dots are present on the back and sides; those on each side often form a distinct dark line, but those on the back do not. The dark herringbone pattern on the sides of the tail is characteristic. The long tail can be almost two-thirds of the total body length. Both sexes may have cirri, but they are more prominent in males.

Larvae and Juveniles   The aquatic, gilled larvae are 1.5–2 inches long and are gray or cream-colored with a clear tail fin. Duration of the larval period depends on water temperature and other environmental conditions; it can be as short as 2 months but is usually about 6 months. Larval periods of more than a year and

top   Long-tailed salamanders typically have black dots on the back and sides, no central stripe, and a long tail.

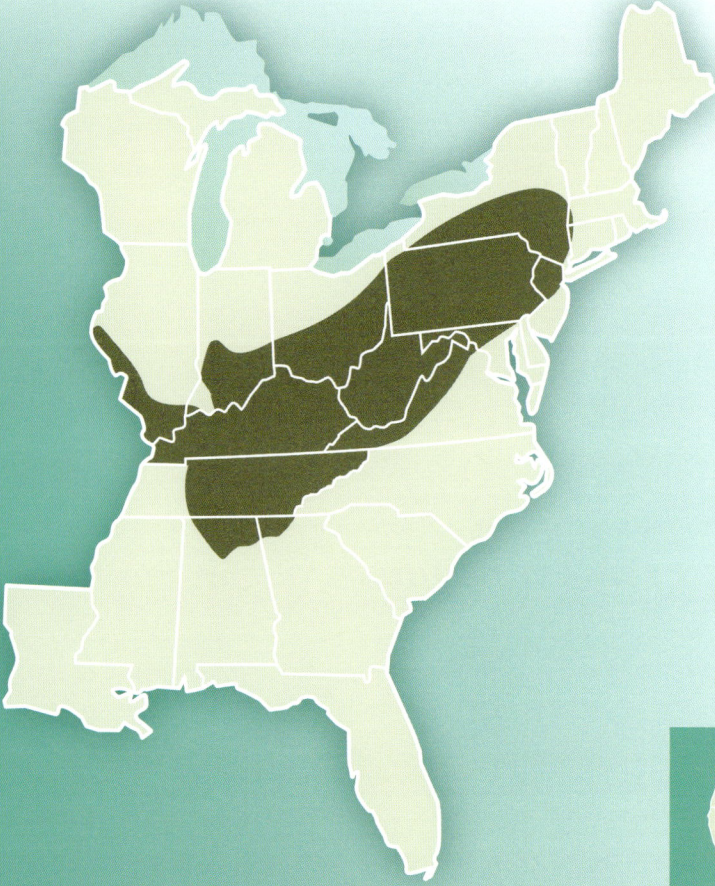

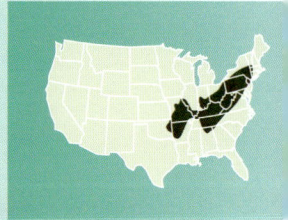

Long-tailed sala-
mander larvae are
aquatic with notice-
able gills.

The background color of long-tailed salamanders can be orangish yellow or brownish yellow.

possibly 2 years have been reported. Juveniles resemble adults, but their black pigment is not as pronounced.

Similar Species The long-tailed salamander is most likely to be confused with the three-lined salamander, which has a distinct stripe down the middle of the back; and the cave salamander, which has a reddish body and scattered black spots on its body and tail. The spots on the tail of the long-tailed salamander, in contrast, coalesce to form a series of Z-shaped transverse bars (herringbone pattern) down the length of the tail.

Distribution The long-tailed salamander's range extends from extreme southern New York in a southwestward direction. It includes Pennsylvania, northern New Jersey, northern Maryland, northern and western Virginia, West Virginia, southern Ohio, southern Indiana, Kentucky, eastern Tennessee, extreme northwest Georgia, northern Alabama to Mississippi, extreme southern Illinois, and

Long-tailed salamanders are mostly terrestrial and are often associated with caves and habitats where shale is present.

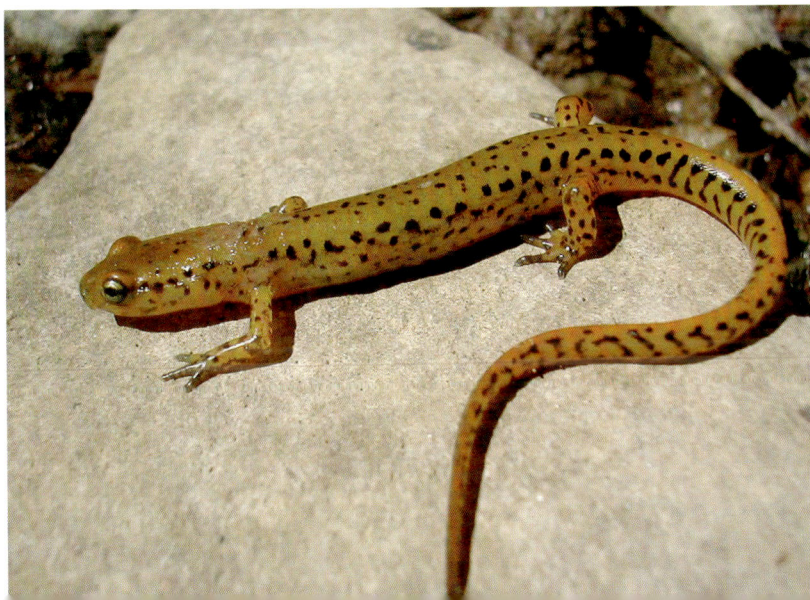

eastern Missouri. The western subspecies (*E. l. melanopleura*) occurs west of the Mississippi River in east-central Missouri, northern Arkansas, and extreme eastern Oklahoma. The range is generally outside that of the three-lined salamander, although some overlap occurs.

Habitat Long-tailed salamanders are mostly terrestrial and are generally associated with seepage areas and the margins of small streams and seasonal wetlands, although individuals are occasionally found in wooded habitats far from water. They typically hide beneath logs, rocks, and other ground cover during the daylight hours and forage at night. The species is especially noted for its association with caves and habitats where shale is present.

Behavior and Activity Long-tailed salamanders associated with surface environments or twilight zones of caves are nocturnal and are especially active on rainy nights during warm periods of the year. Adults readily swim. Congregations of numerous individuals are not uncommon in some areas, suggesting that the species is not territorial. Individuals living in forests migrate to aquatic sites during the breeding season, and populations have been observed migrating in and out of caves and mine shafts.

Reproduction Long-tailed salamanders mate from early fall through winter, with the actual months of reproductive activity determined by geographic location, altitude, and seasonal temperatures and rainfall patterns. Mating is presumably aquatic, and a female may lay from 60 to as many as 100 eggs from fall to spring. The eggs are usually laid singly or in clusters that become scattered, and the female does not remain with the clutch.

Female long-tailed salamanders lay more than 60 eggs, singly or in clusters that become scattered. The female does not remain with the eggs after they are laid.

Terrestrial predators of long-tailed salamanders probably include snakes, larger salamanders, and shrews.

Food and Feeding Adults eat almost every kind of small invertebrate in their habitat: insects such as beetles, flies, and moths; spiders; millipedes; isopods; worms; and springtails. Larvae eat small aquatic invertebrates such as midge and mayfly larvae, ostracods, copepods, and snails.

Predators and Defense Sculpins and sunfish are known aquatic predators of long-tailed salamanders, and terrestrial predators probably include snakes, larger salamanders, and shrews. When threatened, adults run to cover or wave the tail, which breaks off if a predator grabs it.

Conservation The species is widespread and often abundant and does not appear to be seriously threatened on a broad scale. Local populations have probably been reduced or possibly extirpated in situations where strip mining or acid runoff from mine tailings has altered stream habitats.

Taxonomy The long-tailed salamander was described in 1818. Three subspecies were recognized until 1998, when genetic studies revealed the three-lined salamander (*Eurycea guttolineata*) to be a distinct species. The dark-sided salamander (*E. l. melanopleura*) of the Ozark highlands is still considered a valid subspecies.

JUNALUSKA SALAMANDER *Eurycea junaluska*

BODY SHAPE
Small and slender

**NUMBER
OF COSTAL
GROOVES**
14

BODY COLOR
Yellowish brown

PATTERN
Dark brown mottling on back; wavy
stripes on sides

BODY SIZE
Metamorphs about
1.6 inches; adults up
to 4 inches in total
length

Description The body is yellowish brown with small, irregular darker brown spots on the back and sides. The markings on the sides form wavy, disconnected stripes. The belly is light greenish yellow and unmarked. The tail constitutes about half the total length. During the breeding season, males may develop a small mental gland and have short cirri and enlarged head muscles.

Larvae and Juveniles The gilled, aquatic larvae are olive to brown with light spots on the back and sides and a clear tail fin. They are about 0.5 inches long at hatching and about 3 inches long at metamorphosis, which occurs after 2–3 years. The larvae are essentially adult size at metamorphosis.

top Junaluska salamanders have small, irregular dark brown spots on the back and sides that form wavy, disconnected stripes on a yellowish brown body.

Junaluska salamander larvae are gilled, aquatic, and have olive to brown bodies with light spots on the back and sides.

Similar Species The Blue Ridge two-lined salamander, which is found in the same habitat, is more orange than the Junaluska and has distinct stripes partway down the sides. Dusky salamanders have a distinctive light line from the eye to the jaw.

Distribution The geographic range of the Junaluska salamander is restricted to Graham County in western North Carolina and adjacent counties in eastern Tennessee.

The Junaluska salamander is listed as a Threatened species in North Carolina and as Imperiled and in need of management in Tennessee.

Habitat Junaluska salamanders are normally found under rocks and logs along margins of streams bordered by hardwood forest habitat.

Behavior and Activity Adults and juveniles leave their hiding places at night to hunt; adults are often found crossing roads on rainy nights.

Reproduction Males may develop small cirri, mental glands, and enlarged head muscles during the breeding season, but courtship and mating have not been observed. In the spring, females lay clusters of 30–68 eggs (average 38) in streams more than a foot deep, attaching them to the underside of rocks. The female remains with the eggs until they hatch after about a month, usually in April, May, or June. Metamorphosis occurs in May to August.

Food and Feeding The diet is not known but presumably consists of a variety of small aquatic and terrestrial invertebrates.

Predators and Defense No predators have been documented, but larger salamanders, snakes, shrews, and birds would presumably regard these little salamanders as prey.

Conservation The species is susceptible to timbering and other activities that affect stream water quality. Because of their limited geographic distribution and relative rarity in most habitats where they occur, local populations are in danger of being eliminated if their streams and the forests surrounding them are not properly managed. The Junaluska salamander is listed as a Threatened species in North Carolina and as Imperiled and in need of management in Tennessee.

Taxonomy The Junaluska salamander was first recognized as being distinctive from the two-lined salamander in 1937 and was described as a new species in 1976. Molecular research shows that this species is very closely related to the brownback salamander.

CAVE SALAMANDER   *Eurycea lucifuga*

BODY SHAPE	NUMBER OF COSTAL GROOVES	BODY COLOR	BODY SIZE
Slender; wide, flat head; large eyes	14 or 15	Orange	Metamorphs 2.5 inches; adults up to 7.1 inches in total length
		PATTERN Large black spots on body and tail	

Description   These long, slender salamanders have a wide, flat head; large eyes; a long, prehensile tail; and a body strikingly colored in orange, reddish orange, or yellowish orange. The body and tail are covered with large, irregularly placed black spots. The belly is whitish with no dark markings.

Larvae and Juveniles   The aquatic, gilled larvae are about 0.5 inches in total length at hatching and are pale brown with lengthwise rows of light spots on the sides. Juveniles are yellowish but become reddish as they mature.

top   A slender, orange body and wide, flat head with large eyes are trademark traits of the cave salamander.

bottom   Hatchling cave salamanders are pale brown with lengthwise rows of light spots on the sides.

Juvenile cave salamanders (*left*) are yellowish but become reddish as they mature (*right*).

Similar Species The cave salamander is most likely to be confused with the long-tailed salamander, but the spotting on the sides of the tail of the latter coalesces into a herringbone pattern. Long-tailed salamanders are also more yellowish than orange and do not have the flattened head characteristic of cave salamanders.

Distribution Cave salamanders are found from just north of Roanoke, Virginia, southwestward along the border of western Virginia and eastern West Virginia south to northwestern Georgia, and westward to northern Alabama and northeastern Mississippi. They also occur in the eastern three-fourths of Tennessee, in most of Kentucky and into Indiana and southern Illinois, and west to Missouri, Arkansas, and eastern Oklahoma.

Habitat The cave salamander is associated with karst (limestone) topography, and most populations are found in and around caves or rocky cliff habitats, commonly in the twilight zone. Large numbers have been found hundreds of yards inside caves, but individuals are also found in wooded habitats in the vicinity of rock outcrops, caves, and abandoned mine shafts.

Behavior and Activity Cave salamanders are noted for climbing on and within rock faces and crevices, using their prehensile tail for balance. They are often found beneath rocks, logs, leaves, or other ground litter, and may be active on the surface during rains.

DID YOU KNOW?

Unlike frogs, salamanders cannot hear airborne sounds, and most cannot vocalize.

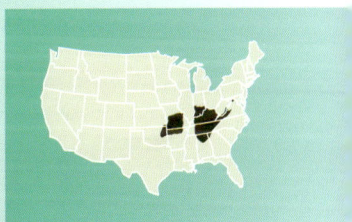

Cirri, which appear during the mating season, are visible on this male cave salamander.

Reproduction Courtship and mating are presumed to occur during summer and early fall, with eggs being laid from September to February, the timing varying among localities. Single eggs are laid in rocky stream pools inside or outside caves. Each egg is attached to the bottom, and a female may lay up to 120 eggs (average more than 60). The larval period is at least a year and may be 15–18 months.

Food and Feeding Cave salamanders are known to prey on more than 100 different kinds of invertebrates, including flies and beetles, springtails, isopods, millipedes, mites, earthworms, snails, spiders, centipedes, crayfish, and snails; they also eat juvenile slimy salamanders.

Predators and Defense Potential predators include other salamanders, snakes, shrews, and birds. Cave salamanders will hide the head beneath the body and wave the tail, which contains toxic compounds and will detach if grabbed by a predator.

Conservation Although cave salamanders are widespread and abundant in some areas, populations that are strongly associated with cave systems are vulnerable if the surrounding forest habitat or water quality of streams within or outside caves is degraded.

Taxonomy The cave salamander was described as a species in 1822. No subspecies are recognized.

Cave salamanders are associated with karst (limestone) topography, and most populations are found in and around caves or rocky cliffs.

CHAMBERLAIN'S DWARF SALAMANDER   *Eurycea chamberlaini*

BODY SHAPE
Small and slender

NUMBER OF COSTAL GROOVES
15 or 16 (usually 16)

BODY COLOR
Yellowish to brown

PATTERN
Dark stripes along the sides

BODY SIZE
Adults up to 3.3 inches in total length

Description Chamberlain's dwarf salamander is a small, slender salamander with a yellowish-brown back bordered by dark stripes along its sides and a yellow belly. Unlike most eastern Plethodontidae (lungless) salamanders, it has only four toes on each back foot.

Chamberlain's dwarf salamander has a yellow belly.

Larvae and Juveniles Hatchlings are less than 0.3 inches in total length with red gills and a transparent tail fin that extends forward in front of the hind

top The presence of only four toes on each back foot is a key character of the Chamberlain's dwarf salamanders.

CHAMBERLAIN'S DWARF SALAMANDER *Eurycea chamberlaini*

legs. Young larvae and young juveniles are grayish brown. The young have spots until they assume the adult coloration.

Similar Species Chamberlain's dwarf salamander can be confused with the southern two-lined salamander, which is larger and has five toes on each back foot. The four-toed salamander has four toes but has a constricted tail base and an enamel white belly with black spots. Any of the other dwarf salamanders could be confused with Chamberlain's dwarf salamander, but with the exception of the dwarf salamander (*Eurycea quadridigitata*), they do not overlap in range. The dwarf salamander and Chamberlain's dwarf salamander only have limited overlap in their ranges in central South Carolina. The dwarf salamander is usually darker brown in coloration whereas Chamberlain's dwarf salamander is yellowish.

Distribution Chamberlain's dwarf salamander is known from the Piedmont of South Carolina and North Carolina, extending into the upper Coastal Plain of South Carolina and central Coastal Plain of North Carolina. An apparent disjunct population in northwestern South Carolina (Anderson and Pickens Counties) may prove to be contiguous with populations further south in Aiken and Barnwell Counties. Contrary to previous published reports, Chamberlain's dwarf salamander is not known from Georgia or Alabama, and no records of this species are currently known from west of the Savannah River.

Habitat Adults are found near streams or ponds in damp woodlands, bottomlands, and swamps.

Behavior and Activity Adults migrate from more terrestrial habitats to wetlands during the fall. They are found under leaves, logs, pine needles, and other surface debris.

Reproduction Adult Chamberlain's dwarf salamanders migrate from November to February, timing their movement with warmer rains above 50°F. Females

This Chamberlain's dwarf salamander is exhibiting its defensive posture.

lay 40–45 eggs, usually attaching then singly to the underside of aquatic vegetation or debris, and have been reported to guard their eggs. Larvae metamorphose in as little as two months or as many as six months.

Food and Feeding These small salamanders eat mites, earthworms, millipedes, small spiders, and a variety of small insects.

Predators and Defense Predators include larger salamander species, semiaquatic snakes, frogs, predaceous insects, and spiders. An antipredator defense behavior observed by one of the authors is curling into a circle and hiding its head beneath its tail.

Conservation This is a secretive species that seems to disappear from observation seasonally. Protection of its wetland habitats from disturbance, drainage, and development is important.

Taxonomy Chamberlain's dwarf salamander was once considered a yellow color morph of the dwarf salamander (E. quadridigitata). It was formally described as a distinct species by Guttman and Harrison in 2003. In 2017, E. quadridigitata was further split into three additional species.

DWARF SALAMANDERS

Dwarf Salamander   *Eurycea quadridigitata*

Western Dwarf Salamander   *Eurycea paludicola*

Hillis's Dwarf Salamander   *Eurycea hillisi*

Bog Dwarf Salamander   *Eurycea sphagnicola*

BODY SHAPE	**NUMBER OF COSTAL GROVES**	**BODY COLOR**	**BODY SIZE**
Tiny and slender	15–18	Yellowish to brown-tan	Metamorphs 0.8 inches; adults up to 5 inches in total length
		PATTERN	
		Dark stripe down each side	

Description These small, slender salamanders have a yellowish to brown or tan back bordered by dark stripes along the sides that continue onto the tail. Dwarf salamanders have a gray belly and a tail that is square-like in cross-section. They differ from most other eastern salamanders in having only four toes on each hind foot. The dwarf salamander (*E. quadridigitata*) has 17 or 18 costal grooves; the western dwarf salamander (*E. paludicola*) has 14–17 costal

top Dwarf salamanders are common in and around many wetland habitats.

grooves and is 1.77–3.4 inches in total length; Hillis's dwarf salamander (E. hillisi) has 14 costal grooves and is 1.1–2.8 inches in total length; and the bog dwarf salamander (E. sphagnicola) has 12–14 costal grooves and is 1.49–2.6 inches in total length, making it the smallest of the dwarf salamanders. Bog dwarf salamanders are variable in dorsal coloration from bronze-cooper to brownish-red or gold. They may be nearly patternless or with a vertebral stripe. The sides have a black to dark gray stripe that is thinnest from the eye to the front legs and then becomes broad and often covers the entire body length. Distinguishing between these species is difficult. Some can be identified based on their geographic range and some by variation in costal grooves, habitats, coloration, or micromorphological differences, but they are primarily separated based on molecular differences.

Larvae and Juveniles Hatchlings are less than 0.3 inches in total length. They have red gills and a transparent tail fin that extends forward in front of the hind legs except for in the bog dwarf salamander where the dorsal tail

top Dwarf salamander

top middle Hillis's dwarf salamander

bottom middle Bog dwarf salamander

bottom Western dwarf salamander

Bog dwarf salamanders occur along the Gulf Coast in Mississippi, Alabama, and the western panhandle of Florida.

fin inserts above the cloaca. Young larvae and young juveniles are grayish brown on top. The young have spots and acquire the adult pattern as they age.

Similar Species Dwarf salamanders could be confused with the southern two-lined salamander, which gets larger and has five rather than four toes on its hind feet. The other salamander that looks similar is the four-toed salamander, which like the dwarf salamander has four toes on the rear feet but has a constricted tail base and an enamel white belly with black spots.

Distribution The distribution of the dwarf salamanders is a bit confused because they were considered a single species for over 150 years. The dwarf salamander (E. *quadridigitata*) is known throughout most of the Coastal Plain from southern North Carolina southward through eastern South Carolina and into southern Georgia and all of Florida. Its range extends west through southwestern Alabama and extreme eastern Louisiana. Although not ver-

The dwarf salamander, Hillis's dwarf salamander (right), and western dwarf salamander leave their forested habitat or the margins of wetland areas for the wetland itself during the breeding season in late fall and early winter. They return to terrestrial habitats in the spring.

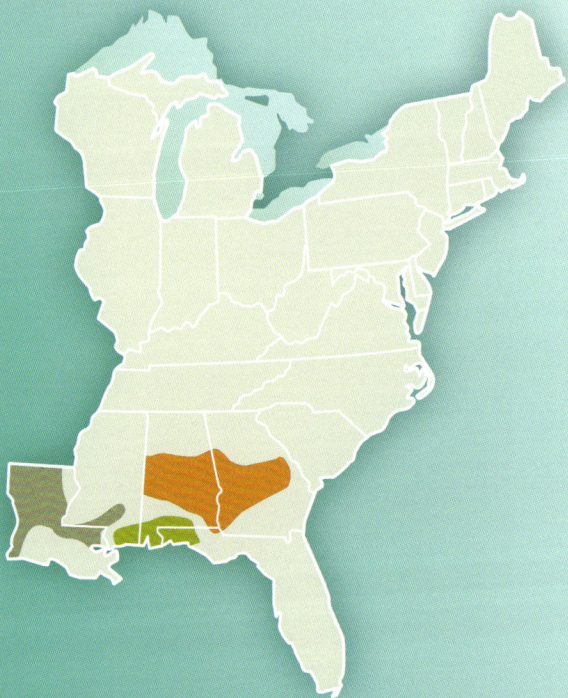

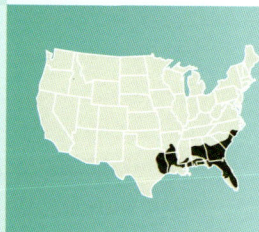

- ● dwarf salamander
  Eurycea quadridigitata
- ● western dwarf salamander
  Eurycea paludicola
- ● Hillis's dwarf salamander
  Eurycea hillisi
- ● bog dwarf salamander
  Eurycea sphagnicola

ified since the splitting of this species in 2017, it might have occurred in the coastal wetlands of southern Alabama. The western dwarf salamander ranges from the eastern third of Texas through most of Louisiana to eastern Mississippi (at least to Jones County). Hillis's dwarf salamander occurs throughout the southern half of Alabama except for Mobile Bay and near the Florida border in the western part of the state; eastward into mid-central Georgia to the western edge of the Ogeechee River Basin; and between the Chipola and Choctawhatchee Rivers of the central Florida panhandle. The bog dwarf salamander is known from within 50 miles of the Gulf Coast in Mississippi, Alabama, and the western panhandle of Florida. The ranges of all five species will undoubtable change and be better delineated as more research is conducted after their definition and redescription in 2017.

Habitat Dwarf salamanders are common in and around many wetland habitats, including Carolina bays, seepage areas, pines in pine flatwoods, sphagnum bogs, and swamp margins. They are sometimes active above ground during the day when they are migrating. The dwarf salamander (E. quadridigitata) often breeds in wetlands associated with cypress trees.

Behavior and Activity With the exception of the bog dwarf salamander, the dwarf salamanders leave their forested habitat or the margins of wetland areas for the wetland proper during the breeding season in late fall and early winter and then return to the more terrestrial habitats in the spring. They are most commonly found among and beneath fallen leaves, logs, pine straw, and other ground cover.

Reproduction Dwarf salamanders move to their more aquatic breeding sites from late summer into the fall. The exact timing varies geographically and in response to seasonal weather conditions. The bog dwarf salamander may not migrate far or at all, as both larvae and adults have been found year-round in their breeding sphagnum mats. Female dwarf salamanders lay 40–45 eggs, usually attaching them singly to the underside of vegetation in the water. Some lay their eggs in low areas without standing water and wait for fall and winter rains. Dwarf salamander are not known to guard their eggs. The larvae metamorphose in as few as 2 months or as many as 6, depending on the rate of water drying.

Food and Feeding They feed upon small insects, spiders, mites, and millipedes.

Predators and Defense In some parts of their range pig frogs and black swamp snakes are known predators, as are a variety of semi-aquatic snakes, frogs, salamanders, predaceous insects, and spiders

Formerly considered a subspecies of E. *quadridigitata*, the western dwarf salamander was elevated to species status in 2017.

Conservation The dwarf salamanders are common in most areas within their ranges but local populations can easily be extirpated if wetlands are drained or otherwise damaged or altered. The recent taxonomic reevaluation of this species complex has identified a need for population and ecological research to better define the geographic ranges, life histories, and conservation needs of all four species.

Taxonomy The dwarf salamander (E. *quadridigitata*) was originally described in 1947 and was revalidated in 2017. The western dwarf salamander was originally described in 1947 as a subspecies of E. *quadridigitata* and was elevated to species status in 2017. Hillis's dwarf salamander and the bog dwarf salamander were described in 2017.

SPRING SALAMANDER   *Gyrinophilus porphyriticus*

BODY SHAPE
Robust; truncated head; tail fins present

NUMBER OF COSTAL GROOVES
17–19

BODY COLOR
Salmon to pink

PATTERN
Small streaks, spots, or chevrons

BODY SIZE
Metamorphs 3–5 inches; adults up to 9 inches in total length

Description Spring salamanders are stout-bodied with irregular black streaks, spots, or reticulation on a salmon to yellowish brown to reddish orange body, head, and tail. A light white to yellow line usually bordered by gray to black extends along the canthus rostralis from the eye to the tip of the broad snout. The belly is pinkish and unmarked. The tail is flattened from side to side. Individuals become darker with age.

Larvae and Juveniles The larvae have a long, truncated snout; external gills; and a flattened tail with fins. Body color varies from pink to light yellowish brown to light gray with black flecking. The larvae become quite large (to 6 inches) before they metamorphose into adults in 4–5 years. They lack the light or dark spots found on the larvae of other salamanders in the region.

top The Blue Ridge spring salamander is a brighter red with fewer spots than the other subspecies of spring salamander.

The yellow line from the eye to the tip of the snout marks the canthus rostralis. White flecking is sometimes evident on the lower jaw of spring salamanders.

Spring salamanders have stout, robust bodies and a flattened tail. Older individuals like this one (top) often become darker in body color.

Similar Species No other salamander in the range of this species has a light line on a rounded snout. Dusky salamanders and related species have a prominent eye-to-jaw stripe. Mud salamanders are reddish brown with scattered distinct, round black dots. Red salamanders are reddish with numerous irregular black dots that run together with age. Tennessee cave salamanders have tiny eyes and a longer head and wider snout than spring salamander larvae.

Distribution Spring salamanders occur in the mountains and foothills from Maine and southern Quebec, Canada, southwest to northern Georgia, Alabama and extreme northwest Mississippi. There are four recognized subspecies with northern spring salamanders having the largest range, occurring from Maine to northwest Georgia, northern Alabama, and extreme northeastern Mississippi. Kentucky spring salamanders are found in eastern Kentucky and southwestern Virginia; Blue Ridge spring salamanders occur in the Blue Ridge region of eastern Tennessee and western North Carolina; and the Carolina spring salamander ranges from southwestern North Carolina to east-central Alabama.

Habitat Spring salamanders inhabit cool streams, springs, and caves in the Appalachian Mountains. They tend to avoid stream reaches with fish and are most abundant in headwater streams and seepage areas. Adults hide under rocks and in debris in streams during the daytime and prowl at night for prey, both in the water and on land. The larvae remain in gravel and under rocks until nightfall. Adults move into the surrounding hardwood forests during wet periods and can be found under logs and rocks. Larvae and adults are commonly found in caves when they are present in an area.

Behavior and Activity Spring salamanders can be found in streams in every month of the year; during winter months, even at high elevations, they move around beneath rock cover and in crevices. They occasionally venture out from the water on rainy nights and can be found on roads and under logs and rocks in the surrounding forest. The larvae may occur in high densities, in part because the larval period is so long.

Reproduction Most of the reproductive activities take place in water, below the surface. Mating takes place in fall and spring, and perhaps in winter in the southern portion of the range. Females lay 16–106 eggs in late spring and summer in thin layers under rocks in water by turning upside down and using their arched back and tail to brace themselves. The eggs hatch in summer and fall. Lowland populations reach maturity a year earlier and at a smaller size (about 3 inches in total length) than high-elevation populations (about 5 inches in total length).

Food and Feeding Spring salamanders are voracious predators of other stream salamanders—including dusky, red, two-lined, and woodland salamanders—and can influence both population sizes and individual growth rates of prey species. Numerous invertebrates such as worms, isopods, spiders, centipedes, beetle larvae, slugs, and snails are also eaten. Head size determines

Spring salamanders are voracious predators of other stream salamanders, such as this Ocoee salamander, as well as of a variety of invertebrates. Adults will even eat larvae and juveniles of their own species.

Spring salamanders, such as this Kentucky spring salamander, occasionally venture away from streams on rainy nights and travel into the surrounding forest.

the size of prey a larva or adult can eat. Adults eat larvae and juveniles of their own species as well.

Predators and Defense Northern watersnakes, garter snakes, and shrews are known predators, and predatory fish probably take these salamanders as well. Adults will tuck their head beneath their body and elevate and wave their tail when attacked by small mammals. Secretions from the tail are noxious to shrews.

Conservation Spring salamanders are listed as Threatened in Mississippi. Siltation caused by deforestation, urbanization, agricultural activities, and construction above headwater streams can cause spring salamander populations to decline.

Taxonomy This species was first described in 1827. Four subspecies are recognized: the northern spring salamander (*G. p. porphyriticus*); Blue Ridge spring salamander (*G. p. danielsi*), described in 1901; Carolina spring salamander (*G. p. dunni*), described in 1941; and Kentucky spring salamander (*G. p. duryi*), described in 1930.

MUD SALAMANDER   *Pseudotriton montanus*

BODY SHAPE
Robust body; short,
thick tail

**NUMBER
OF COSTAL
GROOVES**
16 or 17

BODY COLOR
Reddish to brownish

PATTERN
Distinct round,
black spots on body
and tail

BODY SIZE
Metamorphs 1.6
inches; adults up
to 8 inches in total
length

Description Mud salamanders are stout-bodied and blunt-headed. Body size varies geographically (total length reaches 8 inches in the eastern subspecies, *P. m. montanus*), and females get larger than males. The body color ranges from bright red or pinkish to reddish brown depending on the subspecies, local variation, and age and size of the individual. Well-defined, round black spots are scattered on the body and tail, although spotting may be reduced or absent in some forms. The irises are brown. The midland mud salamander (*P. m. diastictus*) is more brightly colored than the other subspecies, and the rusty mud salamander (*P. m. floridanus*) lacks distinct spots on the body. The body length of *P. m. montanus* averages about 1.5 times that of the other subspecies.

top Mud salamanders, such as this midland mud salamander, typically are stout bodied and have a blunt head and brown irises. The body and tail are usually red with well-defined, round black spots.

Larvae and Juveniles The gilled, aquatic larvae are less than 0.5 inches long at hatching and average about 1.5 inches at metamorphosis. They are pale brown with dark spots on the back and dark markings on the sides that form streaks. The larval stage lasts more than a year; most individuals metamorphose after 15–17 months. Juveniles resemble adults in color and spotting pattern.

The larvae of mud salamanders are aquatic with distinct gills.

Similar Species The mud salamander is most likely to be confused with the red salamander. The body color and ill-defined spotting patterns are highly variable in both species, but the red salamander has yellowish eyes and the mud salamander has solid brown eyes. In areas where spring salamanders overlap with mud salamanders, the latter are recognizable in not having a distinctive line from the eye to the tip of the snout.

Distribution Mud salamanders occur from New Jersey south to northern Florida and west through southern Mississippi to easternmost Louisiana. An isolated population is known from east-central and northeastern Mississippi, but the species is absent from the rest of northern and central Mississippi and from the western and northern halves of Alabama and from most of north-

Red salamanders (*top*) and mud salamanders (*bottom*) are often confused, but mud salamanders usually have fewer spots, can be duller red in color, and have a shorter, rounder snout.

ern Georgia. The midland mud salamander has a distinct range that includes southern Ohio, western West Virginia, western Virginia, eastern Kentucky, and northeastern Tennessee.

Habitat Mud salamanders typically live in wet areas with muck and mud, particularly those with sphagnum moss and a decaying leaf layer in hardwood forests. The source of water is variable and includes seepage areas below bluffs, small streams, and floodplain swamps.

Behavior and Activity Mud salamanders usually stay hidden, often in burrows in muddy, mucky sites; or beneath dead leaves, logs, and moss in wet habitats. They may move short distances over land during the breeding season, but no well-defined seasonal activity patterns are apparent.

MUD SALAMANDER *Pseudotriton montanus*

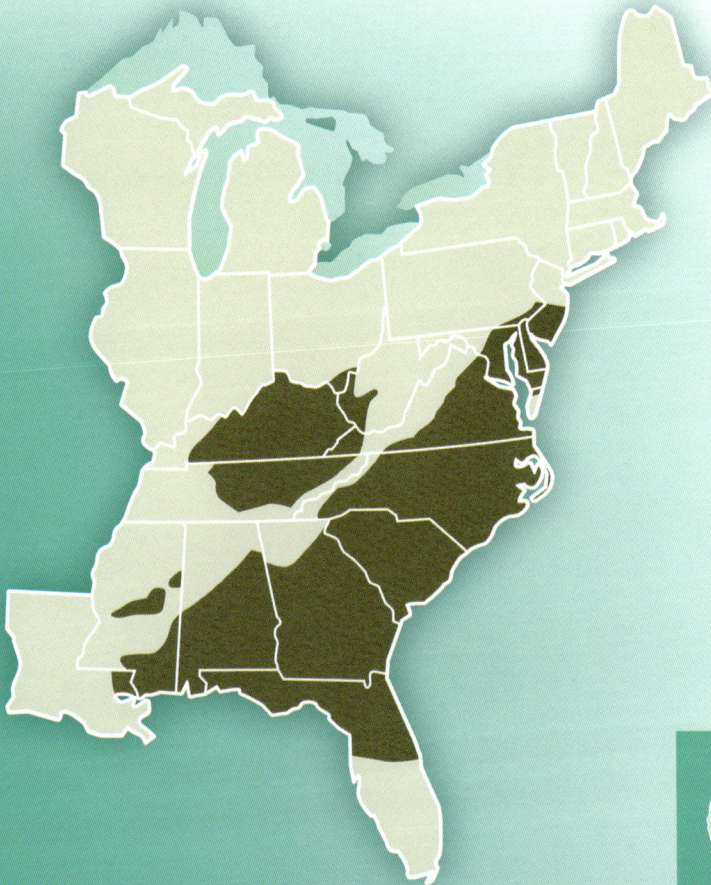

above Mud salamanders typically
live in wet areas with muck and mud,
particularly those with sphagnum
moss and a decaying leaf layer.

right Mud salamanders may move
short distances overland during the
breeding season, but no well-
defined seasonal activity patterns
are apparent.

Reproduction Mud salamanders mate in late summer or fall and probably lay
eggs by December or January that hatch by February or March. Courtship has
not been observed. Females in some South Carolina populations have been
noted to carry 77–192 eggs, although lower numbers may be more common
in other geographic regions. Actual clutch sizes above 30 eggs have not been
found. The eggs are attached to the underside of moss, roots, or decaying vegetation in wet areas.

Food and Feeding Adult mud salamanders are known to eat two-lined salamanders and worms. They presumably will eat other types of salamanders,
small insects, and other invertebrates. The larvae eat aquatic invertebrates.

The rusty mud salamander (*left*) and Gulf Coast mud salamander (*below*) are two of the recognized subspecies of mud salamander.

Predators and Defense Northern watersnakes and eastern garter snakes eat mud salamanders; larger salamanders, ringneck snakes, shrews, and birds are other likely predators. When threatened, mud salamanders curl up so that the head is hidden beneath the body and wave the tail above the body. They produce skin secretions that are at least mildly toxic to some potential predators. The red body color may mimic that of other unpalatable red salamanders and thereby discourage attack by predators, such as birds, with color vision.

Conservation No specific conservation issues have been targeted for this species, although local populations are likely to be affected by habitat degradation that affects stream systems and seepage areas.

Taxonomy The mud salamander was first described in 1849. Four subspecies are recognized: the eastern mud salamander (*P. m. montanus*, described in 1849), midland mud salamander (*P. m. diastictus*, described in 1941), Gulf Coast mud salamander (*P. m. flavissimus*, described in 1856), and rusty mud salamander (*P. m. floridanus*, described in 1942).

RED SALAMANDER   *Pseudotriton ruber*

BODY SHAPE	NUMBER OF COSTAL GROOVES	PATTERN	BODY SIZE
Robust body with thick tail	16 or 17	Black spots irregularly scattered over the entire body	Metamorphs 2 inches; adults up to 7.1 inches in total length
	BODY COLOR Reddish to orangish or purplish		

Description   Red salamanders are robust and have a thick tail. Females, which are larger than males, commonly exceed 6 inches in total length. The body and tail are red, but the shade ranges from deep or bright red to orangish, purplish, or brownish depending on geographic variation and the size and age of the individual. Black spotting is present on most individuals; the spots are not distinct and often fuse together, especially on older animals, causing them to look purplish. The belly is pink to red and in adults usually has small black spots. The irises are yellow. Black-chinned red salamanders have heavy black pigment on the chin and lower lip; Blue Ridge red salamanders have little to no dark pigment on the last half of the tail and none on the chin and lower lip; and southern red salamanders are usually covered in white flecks, especially

top Red salamanders, such as this northern red salamander from West Virginia, range from deep or bright red to orangish, purplish, or brownish, and most individuals have black spotting.

Hatchling red salamanders are about 0.5 inch in body length, and the aquatic larvae have red gills. The larval stage in most populations lasts at least 1.5 years and can last more than 3 years in some.

on the head and sides, in addition to the black spots.

Larvae and Juveniles The hatchlings are about 0.5 inches in body length. The aquatic larvae have red gills; a pale brown body, sometimes with mottling on the back and sides; and a whitish belly. They metamorphose at about 2 inches in May to July after a growth period of 27–31 months. The larval stage in most populations lasts at least 1.5 years and can last more than 3 years in some. Juveniles resemble adults in color, but the belly lacks spots.

Similar Species Red salamanders can usually be distinguished from mud salamanders, which may have a similar body color and pattern, by the yellow irises; mud salamanders have brown irises. In addition, the length of the snout from the eye to the nose tip is shorter in mud salamanders than in red salamanders. In areas of overlap with spring salaman-

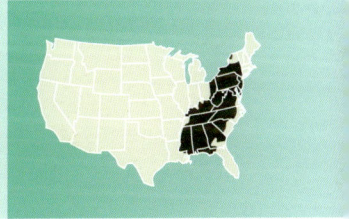

ders, red salamanders can be differentiated by the lack of a noticeable line from the eye to the tip of the snout.

Distribution Red salamanders are found from southern New York, Pennsylvania, eastern Ohio, and southeastern Indiana through part of each southeastern state. They are absent from most of Louisiana and the Mississippi River floodplain of Mississippi, Tennessee, and Kentucky, as well as from most of the Atlantic Coastal Plain and the Florida peninsula.

opposite Red salamanders are associated with hardwood forests and with wet areas, including those caused by natural seeps.

As they age, red salamanders often become less red and have more, less-defined black spotting.

Habitat Red salamanders are associated with hardwood forests and with wet areas caused by natural seeps, stream systems, floodplains, and streams flowing through caves. The breeding and larval habitat is generally slow-moving, clear water with extensive decaying vegetation.

Behavior and Activity Red salamanders stay beneath dead leaves, logs, and other ground litter or in burrows during the day, and may be active on the surface at night. They can be found in all seasons in and at the edge of water in springs, seeps, and small streams, as well as in terrestrial habitats. During spring and fall rains they may move long distances (more than 400 feet) over land between breeding sites and the terrestrial habitats where they spend most of the year. Individuals may be found under logs a long distance from water and are sometimes encountered on wet roads at night. During extreme cold periods red salamanders may go deeper into burrows or ground debris and become inactive.

Reproduction Red salamanders mate in summer or fall following courtship activity, and eggs are laid in fall or winter. The timing depends on such factors as geographic location and seasonal temperature and rainfall patterns. The number of eggs in a clutch ranges from as few as 29 to as many as 130. Some females attach the eggs to the underside of a log or rock and remain with them, sometimes creating and staying in a small depression beneath the eggs that fills with water. Some females lay their eggs deep in the banks of springs and small streams. The eggs hatch into aquatic larvae in 2–3 months.

Food and Feeding Red salamanders eat smaller salamanders as well as worms, insects, snails, spiders, and millipedes. In water, the larvae and adults eat aquatic insects and their larvae.

Predators and Defense Known predators include copperheads; watersnakes, garter snakes, ringneck snakes, shrews, raccoons, birds, and possibly larger salamanders are potential predators. When threatened, red salamanders exhibit elaborate body contortions by hiding the head and waving the tail slowly in the air. Their skin secretions are mildly toxic to some would-be predators. The red body color may mimic that of other red-colored salamanders that are mildly or strongly toxic (such as efts) and might therefore deter attacks by bird predators with good color vision.

There are four subspecies of the colorful red salamander: the northern red salamander (*see photo on page 245*), the southern red salamander (*right*), the black-chinned red salamander (*below left*), and the Blue Ridge red salamander (*below right*).

Conservation The environmental health of this species overall appears to be satisfactory. On a local scale, populations suffer when timbering operations, mining, highway construction, urbanization, or other activities create excessive siltation, acid runoff, or other conditions that degrade the aquatic systems where these salamanders live. Loss of hardwood forests around streams is also detrimental.

Taxonomy The red salamander was described in 1801. Four subspecies are recognized: the northern red salamander (*P. r. ruber*, described in 1801), black-chinned red salamander (*P. r. schencki*, described in 1912), Blue Ridge red salamander (*P. r. nitidus*, described in 1920), and southern red salamander (*P. r. vioscai*, described in 1928).

PATCH-NOSED SALAMANDER   *Urspelerpes brucei*

BODY SHAPE
tiny and slender

**NUMBER
OF COSTAL
GROOVES**
15–16

BODY COLOR
Yellow to yellowish
brown with a light
colored spot at the
tip of the snout

PATTERN
Males with two dark
stripes on the back;
females with no
stripes

BODY SIZE
Adults up to 2 inches
in total length

Description   Adult patch-nosed salamanders are the smallest salamander species in the United States and one of the smallest species in the world. They have
a distinctive saffron yellow patch on the upper surface of the snout and a thin,
distinct yellow line along the midline of the tail. The belly is uniformly yellow.
Males have two distinct dark dorsolateral stripes and a yellow body. Females
are drab and lack the dark stripes but show no significant difference in size
from the males. An indistinct dark line runs posteriorly from in front of the
eye to near the front limbs on each side. Reproductively active males have two
prominent cirri and a mental gland. Adults are small—approximately the size
of newly metamorphosed two-lined salamanders. Each rear foot has five toes,
and the tail is as long as the body.

top The male patch-nosed salamander (*right*) has two dark dorsolateral stripes and a yellow
body whereas the female (*left*) is drab and lacks the dark stripes.

Larvae and Juveniles The upper surface of the snout of larvae has a distinctive white patch. The body and tail are covered densely with small, brownish melanophores except on the snout and the white stripe on the tail. Larvae of the patch-nosed salamander metamorphose and reach sexual maturity at the same time, usually in their third year in the fall (but at a minimum of 2 years). They seem to forego

The patch-nosed salamander has a prolonged larval stage and does very little additional growing after transforming into an adult.

any juvenile stage of their life. Newly metamorphosed individuals are basically full grown with a well-developed skeleton compared to most salamanders.

Similar Species Patch-nosed salamanders resemble dwarf and two-lined salamanders. Dwarf salamanders have four toes on the rear feet, and the tail of both dwarf and two-lined salamanders is much longer than the body.

Distribution Patch-nosed salamanders occur in one small area (maybe as small as 11.5 square miles) in the Appalachian foothills of northeastern Georgia (Stephens and Habersham Counties) and across the Tugaloo River in northwestern South Carolina (Oconee County). A survey conducted in 2021 identified only 25 Georgia localities and 3 South Carolina sites.

Habitat Little is known about this recently described salamander. Adults apparently prefer cover under rocks and in leaf litter on the moist banks of small first-order streams. The tiny larvae live in shallow water in the sand-gravel interstices. Current information indicates that adults spend most of their time underground.

Behavior and Activity Very little is known about the behavior and seasonal activity patterns of patch-nosed salamanders.

Reproduction All of the three adult females known were gravid, with enlarged ova visible through a translucent belly; two had 14 enlarged ova, and one had 6. Adult males collected between April 15 and May 29 had pronounced nasal cirri and enlarged mental glands, suggesting that the breeding season at least covers these two months.

Food and Feeding The structure of the mouth and tongue of adults suggested to the original describers of this species that they capture small insects and other invertebrate prey out of water.

Predators and Defense No predators of patch-nosed salamanders have been described, but their small size makes them very vulnerable to predation by fish,

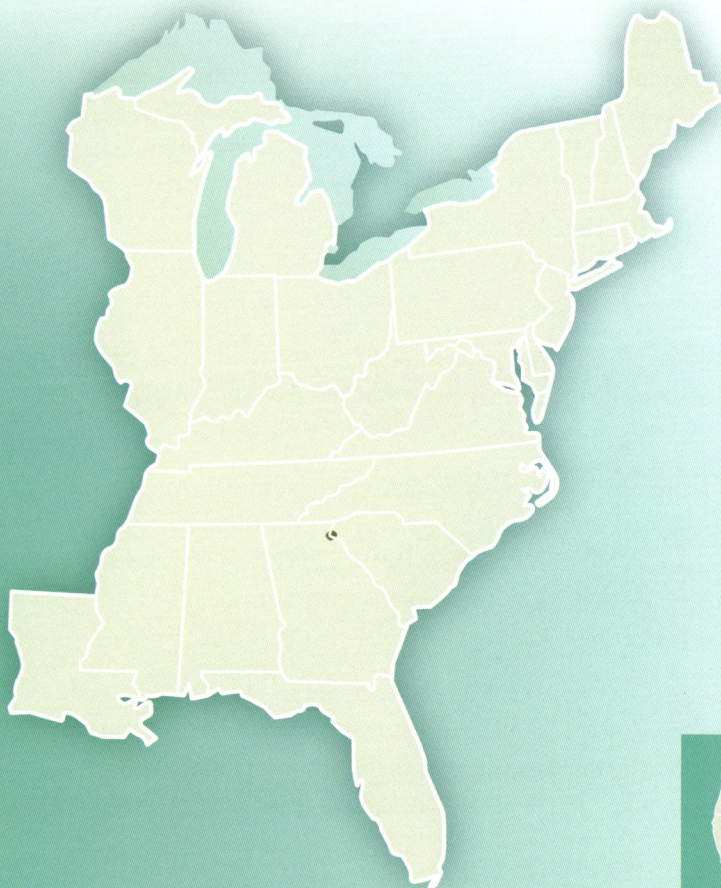

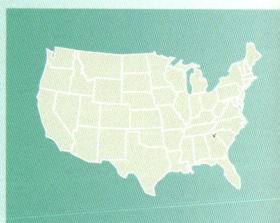

crayfish, and other salamander species. Their vulnerability may be why they seek very shallow water at spring heads and steep-sided ravines. Nothing is known about their defensive behavior.

Conservation This species appears to be extremely rare and may occur in such small numbers as to be in danger of extinction. Georgia lists the patch-nosed salamander as a Species of Special Concern but affords it no legal status.

Taxonomy Patch-nosed salamanders were first described in 2009 by Carlos Camp and coauthors. The new salamander genus (*Urspelerpes*) is the first described in the United States since *Phaeognathus* in 1961.

seasonal wetland
salamanders

previous page A spotted salamander

FLATWOODS SALAMANDERS

Frosted Flatwoods Salamander   *Ambystoma cingulatum*

Reticulated Flatwoods Salamander   *Ambystoma bishopi*

BODY SHAPE
Slender with small head

NUMBER OF COSTAL GROOVES
13–16

BODY COLOR
Brown to black

PATTERN
Frosted with reticulate or irregular specks

BODY SIZE
Metamorphs 2.75 inches; adults up to 5 inches in total length

Description   Frosted flatwoods salamander adults are brown to black with a reticulate pattern of gray to silver "frosting" on the back, head, and tail. The frosting may be irregular or may form a pattern that varies from narrow light lines to netlike. The frosting is concentrated on the lower sides of the body. The head is small and barely larger than the neck. The belly is dark gray with discrete light gray specks or white spots. Males in breeding condition have slightly swollen cloacal glands, but these do not get as large as in other mole salamanders. Reticulated flatwoods salamanders have narrow, orange-gray to

top A mature adult frosted flatwoods salamander of the uniform grayish brown phase. Note the short snout and robust body.

top Frosted flatwoods salamander larvae have a prominent yellow to light tan stripe above the black stripe on the sides of the body and tail. The lower light stripe is broken up in this individual.

bottom Metamorphic reticulated flatwoods salamanders acquire adult coloration and pattern as they grow and mature. The scattered lichen-like markings that cover the body of this metamorph will fade with age.

silvery lines on a chocolate-black background; the belly has a salt-and-pepper pattern with tiny gray flecks.

Larvae and Juveniles Larvae are aquatic and have bushy, reddish brown gills. The dorsal tail fins extend almost to the head. Young larvae have a yellow to gold stripe along the back with a black zone on each side. The yellow stripe changes to tan on older larvae, which have two yellow to light tan stripes on each side of the body and tail; the stripe along the midbody is most prominent. The stripe on the side extends from the mouth and narrows as it passes along the tail. A dark stripe extends from the snout through the eye to the gills, and some older larvae have a black line along the mouth. The belly is cream to white, and the tail fin may have brown mottling. The period from hatching to metamorphosis into terrestrial juveniles is about 12 weeks. Juveniles may retain the tan stripes for nearly a year.

Similar Species Mole salamanders are short and chunky with a large head, and the body is solid gray or has light gray flecking. Spotted salamander juveniles are gray with fine black specks on the back but not on the tail. Juvenile marbled

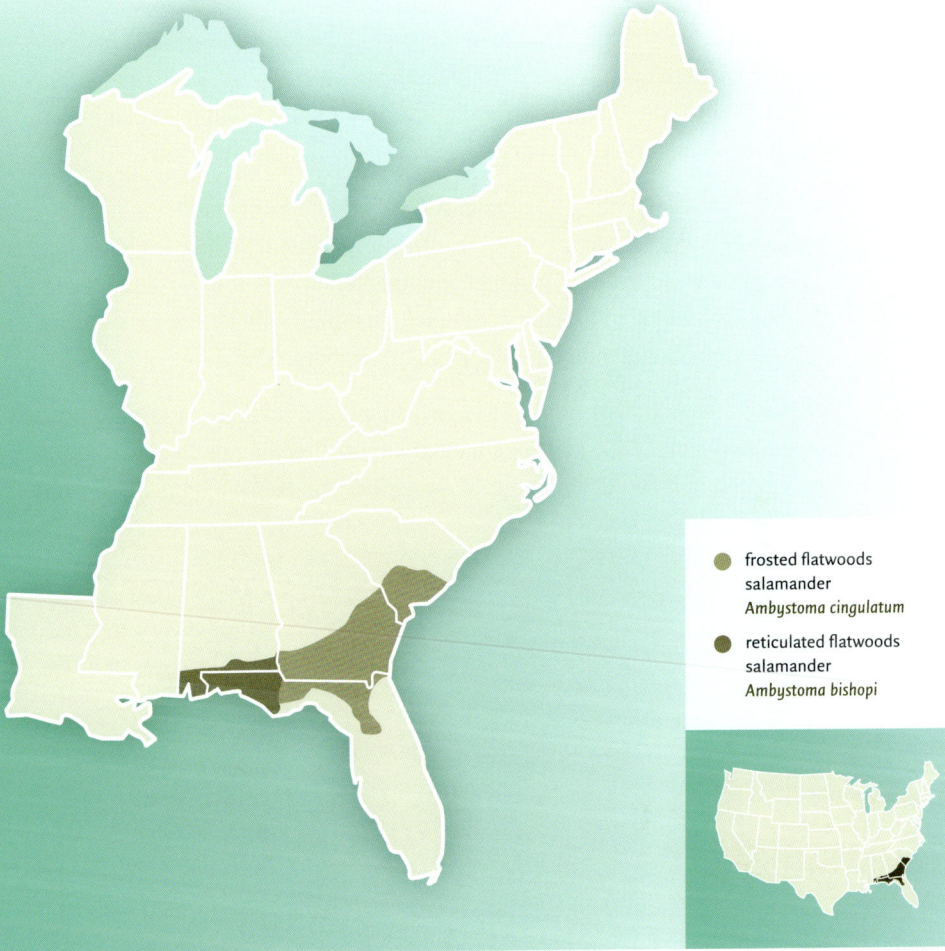

frosted flatwoods
salamander
Ambystoma cingulatum

reticulated flatwoods
salamander
Ambystoma bishopi

salamanders are brown to black with tiny gray to silver specks that may occur in a lichen-like pattern on the back.

Distribution Flatwoods salamanders historically occurred in scattered locations from the central portion of the South Carolina Coastal Plain south to Alachua County, Florida, and west through Georgia to the Florida panhandle to southern Alabama. Populations east of the Apalachicola River are frosted flatwoods salamanders (A. *cingulatum*); those west of the river are reticulated flatwoods salamanders (*A. bishopi*). Recent surveys indicate both species occupy a fraction of their historical range. Frosted flatwoods salamanders have recently been verified at only about 40 sites, one in Georgia (Fort Stewart) and the remainder in Florida, including the St. Marks National Wildlife Refuge, the Apalachicola National Forest, and Flint Rock Wildlife Management Area. The reticulated flatwoods salamander has been verified at about 11 sites within Elgin Air Force Base and 6 within the Escribano Point Wildlife Management Area, all within Florida.

Habitat Adults of both species remain underground most of the year, presumably in rodent burrows and root tunnels in slash pine flatlands and longleaf pine forests. Where the juveniles go in the upland habitat after they leave the breeding ponds is unknown. Breeding sites include cypress and blackgum swamps, roadside ditches, ephemeral pools, and shallow borrow pits.

Behavior and Activity Adults emerge from upland refugia on rainy nights in October and November, mate and lay eggs in the breeding pond, and migrate back to their burrows during rains in December and January. Eggs hatch from

left An adult reticulated flatwoods salamander with profuse silvery frosting and narrow reticulations across the body and tail.

below An adult frosted flatwoods salamander with a somewhat typical pattern of lichen-like markings on a nearly black body.

The dark pattern of brown mottling on the tail is likely used to distract predators so that they attack the tail instead of the head and body.

November to January, and the larvae remain in the pond until March or April. The larvae stay hidden among aquatic plants in shallow water during the day and forage in the open water column at night. After metamorphosis, juveniles migrate into the surrounding uplands and presumably remain underground until they reach maturity.

Reproduction   Courtship and mating take place on land as males and females are approaching breeding ponds. Females lay eggs in small groups of 1–4 eggs in the breeding pond basin, usually under logs and other surface objects that will become flooded during winter rains; or they attach the eggs to vegetation in standing water. Clutch size ranges from about 100 to 225 eggs.

Food and Feeding   Adults and juveniles eat earthworms and small slugs. The aquatic larvae eat small invertebrates and zooplankton.

Predators and Defense   Specific predators are unknown, but aquatic invertebrates such as dragonfly nymphs, predaceous diving beetle larvae, and giant water bugs probably eat the larvae and even adults that remain in the aquatic habitat during warm weather. Large larvae may be cannibalistic. The dark pattern of brown mottling on the tail is likely used to distract predators so that they attack the tail instead of the head and body. Adults will coil the body and wrap the tail around the head when threatened.

Conservation Frosted flatwoods salamanders are listed as Threatened by the U.S. Fish and Wildlife Service (1999) as the result of habitat loss, disease, the small number of currently known locations, and the apparent dramatic decline in numbers at known breeding sites. Alabama, Florida, Georgia, and South Carolina also protect this species either as a Species of Concern or by listing it as Endangered. The reticulated flatwoods salamander is protected under the Endangered Species Act as Endangered (2009) because it faces the same threats faced by the frosted flatwoods salamander. Both of these species have suffered a dramatic decline in numbers and viable breeding sites. It is estimated that their numbers have declined 90 percent since 2000, and breeding sites have declined or disappeared over most of its range. There may not be viable populations left in South Carolina and perhaps only one locality left in Georgia.

Taxonomy The frosted flatwoods salamander was first described in 1868. The reticulated flatwoods salamander was described as a subspecies in 1950 and recognized as a full species in 2006. No subspecies are recognized for either species.

JEFFERSON SALAMANDER *Ambystoma jeffersonianum*

BODY SHAPE	NUMBER OF COSTAL GROOVES	PATTERN	BODY SIZE
Long and robust; broad head	12 or 13	Uniform or with fine silver flecking concentrated on the sides	Metamorphs 2 inches; adults up to 7 inches in total length
	BODY COLOR		
	Bluish gray		

Description Jefferson salamander adults are larger and have a broader head and more elongated snout than other mole salamanders. They are light to dark bluish gray with silver flecking on the lower sides of the body, tail, and limbs. Some adults lack the flecking altogether, and some have silver flecking on the back.

Larvae and Juveniles Larvae are olive green to gray with faded yellow markings on the upper tail fin and on the sides of the head and neck. The belly is silver to white and unmarked. Meta-

Jefferson salamander larvae have large gills and are olive green to gray with faded yellow markings on the upper fin and on the sides of the head and neck.

top Adult Jefferson salamanders are typically uniformly colored without markings, but many retain remnants of silvery flecking and spots.

Newly emerged metamorph Jefferson salamanders are typically thin bodied with a large head.

morphs are uniformly gray with faded yellow spots on the sides and an unmarked belly.

Similar Species Marbled salamander juveniles are brown to black with tiny gray to silver specks that may occur as a lichen-like pattern or as paired faded yellowish spots on the back. Spotted salamander juveniles are olive to gray with fine gray specks on the back but not on the tail. Immature small-mouthed salamanders are brownish gray to black with clusters of light specks or a

JEFFERSON SALAMANDER *Ambystoma jeffersonianum*

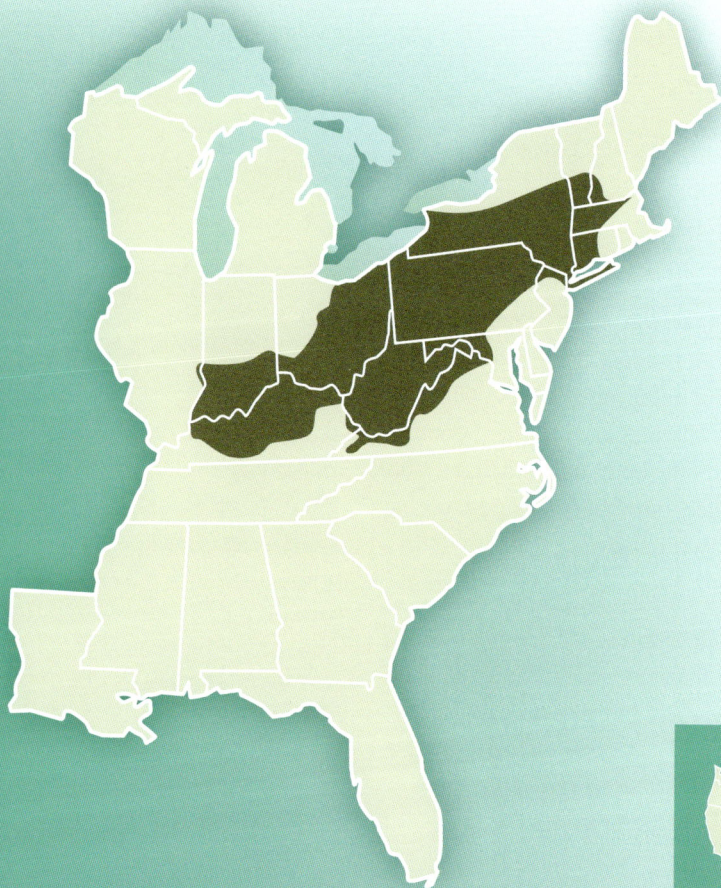

above The yellow spotting of metamorphs changes to silvery spots and flecking in immature salamanders, but these fade with age.

right Adult Jefferson salamanders are larger and have a broader head and more elongated snout than other mole salamanders.

lichen-like pattern mostly concentrated on the sides. Newly metamorphosed juvenile streamside salamanders are uniformly brown to brownish gray with little or no pattern. Blue-spotted salamanders are smaller, profusely blue-spotted, and have shorter toes and a black belly. However, so-called hybrids occur between blue-spotted salamanders and Jefferson salamanders, making some populations almost impossible to distinguish without genetic examination.

Distribution Jefferson salamanders occupy high elevations in the Appalachian Mountains and northern latitudes from southwestern Virginia and central Kentucky northeastward to Ontario, Canada, and New Hampshire.

Habitat These are primarily salamanders of well-drained hardwood forests in montane regions. Adults and juveniles occupy rodent burrows and root tunnels.

Behavior and Activity Adults emerge from underground refugia in late fall before the ground freezes and in winter as soon as the snow melts and the ground thaws, and sometimes before that. They migrate long distances—up to a mile—to reach their woodland breeding pools. Breeding sites may be ephemeral or permanent but are usually devoid of fish, which are major predators. Adults leave the ponds 3–4 weeks after breeding and laying eggs and migrate back to their underground retreats in the forest. They are seldom seen outside the breeding period because they are usually underground.

Reproduction Males usually arrive at the breeding pond before females and lay numerous spermatophores on the pool bottom. Once females arrive, a male will entice a female by behavioral and chemical communication to walk over and pick up a spermatophore with her cloaca. Fertilization of the 140–200 eggs is internal. Eggs are laid 1–2 days later in several gelatinous masses that the female attaches to twigs or other vegetation. The larval period lasts 2–4 months during the summer. Larvae remain hidden under leaves during the day and forage in the open water column at night. Metamorphs move as far as a third of a mile into forest habitat on land and remain underground until they mature 2–3 years later.

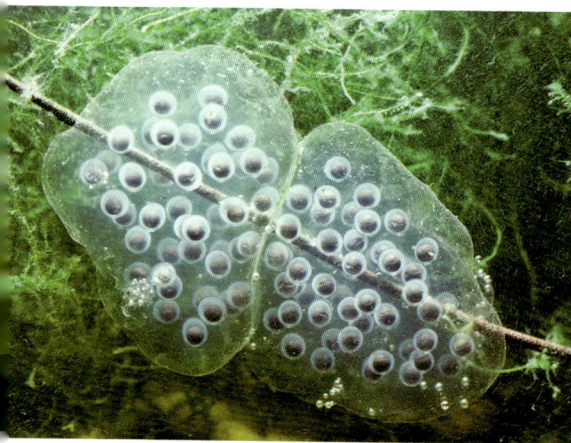

Jefferson salamanders lay their eggs in clusters attached to stems and grasses in water.

Food and Feeding Larval Jefferson salamanders eat a wide variety of aquatic invertebrates and zooplankton. They will also eat the larvae of other salamander species, as well as each other. Adults are thought to eat invertebrates such as earthworms while in terrestrial burrows.

Predators and Defense Striped skunks, raccoons, and shrews are known predators of adults; owls and snakes are suspected predators. Adults elevate the tail, coil the body, hide the head under the tail, and produce a noxious secretion from tail glands that deters potential predators. The larvae are eaten by aquatic invertebrate predators and sometimes by other Jefferson salamanders. The tail fins of larvae become dark in the presence of aquatic invertebrate predators and presumably serve as a decoy that attracts the predator to the tail rather than the head.

Conservation Timber harvest and loss of forest habitat threaten many populations in Kentucky and Virginia. Acid precipitation that lowers pH in the breeding

Jefferson salamanders are found primarily in well-drained hardwood forests. Timber harvest and loss of forest habitat threaten many populations in Kentucky and Virginia.

ponds is another human-made source of mortality. Maintaining mature forests and several breeding ponds in the landscape will be necessary for long-term survival of Jefferson salamander populations.

Taxonomy The Jefferson salamander was first described in 1827. No subspecies are recognized, although in the northern portion of its range this species hybridizes with blue-spotted salamanders to form triploid offspring that are all females (see account on unisexual mole salamanders).

BLUE-SPOTTED SALAMANDER *Ambystoma laterale*

BODY SHAPE
Small and slender;
short legs

**NUMBER
OF COSTAL
GROOVES**
13

BODY COLOR
Dark gray to gray-
black

PATTERN
Large bluish-white
blotches or flecks

BODY SIZE
Adults up to 5.1
inches in total
length

Description Blue-spotted salamanders are small and slender and have dark gray to gray-black bodieswith short legs. Adults have large, bluish-white blotches or flecks on the dorsum and a dark gray venter with small light spots. Females are usually larger with a longer tail than males.

Larvae and Juveniles Hatchlings/larvae are 0.31–0.4 inches in total length and have a dark brown dorsum with faint yellowish transverse bars on the sides. The larvae are typical pond-type larvae with relatively large feathery gills and a large head in proportion to the rest of the body.

Similar Species This species can be confused with unisexual hybrids of blue-spotted and Jefferson salamanders with triploid A. *laterale*–dominated genomes and with other unisexual *Ambystoma*. Positive identification requires

top A young blue-spotted salamander from Long Island, New York

An adult blue-spotted salamander from Maine

looking at cell size, molecular evidence, and/or chromosome number where sympatric unisexual populations exist (see account on unisexual mole salamanders).

Distribution Blue-spotted salamanders occur in glaciated regions of southern Canada and the extreme northern United States, including northeastern Minnesota; all of Wisconsin and Michigan; extreme northern Illinois and Indiana; extreme northwestern Ohio; northern and eastern New York; northern New Jersey and Connecticut; most of Massachusetts; and all of Vermont, New Hampshire, and Maine. There is an isolated population in east-central Iowa. In Canada, the range covers south to the Great Lakes and north to Hudson Bay plus most of maritime Canada extending far to the north.

Habitat Blue-spotted salamanders inhabit primarily deciduous forests or mixed deciduous-coniferous forests but are usually absent from open prairie

A colorful male blue-spotted salamander from northwestern Ohio

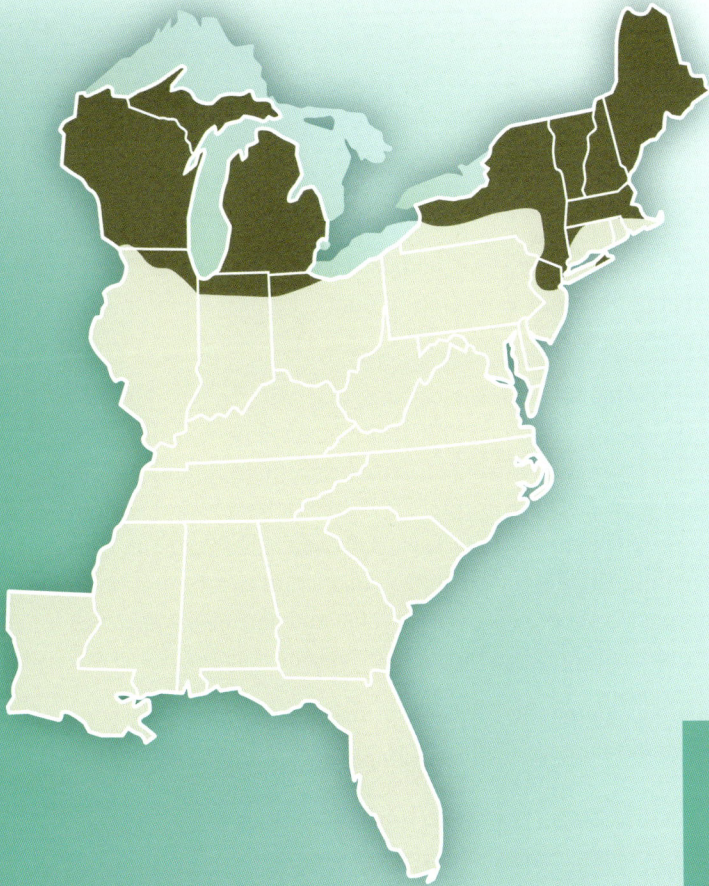

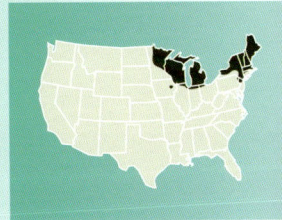

habitats. They do not seem to be adept at burrowing and are often associated with sandy or loamy soils that make burrowing easy.

Behavior and Activity Unlike many of the southern species in the genus *Ambystoma*, the blue-spotted salamander can be active at or near the surface during warmer months of the year. These salamanders require breeding ponds with no fish. Blue-spotted salamanders migrate into the breeding ponds on rainy nights from late winter to early spring. Males arrive before the females.

Reproduction Males and females migrate into fish-free vernal breeding pools, springs, marshes, and roadside ditches during warm rains (45–50°F) in late

A male blue-spotted salamander in its defensive posture

winter to early spring. The timing of migration varies, with some northern populations delaying until as late as May. Soon after courtship, females attach eggs singly or in loosely defined masses of 2–10 eggs to leaves, twigs, detritus, or rocks. Egg counts can sometimes be higher. Embryos hatch three to four weeks after deposition depending upon water temperatures.

Food and Feeding Larvae feed on a variety of aquatic animals with zooplankton and fly larvae probably forming the bulk of their diet. The adult diet varies across the significant north-to-south range. Blue-spotted salamanders are opportunistic feeders, eating snails, earthworms, isopods, centipedes, spiders, insects such as beetles, beetle larvae, springtail nymphs, and slugs.

Predators and Defense Predators of the eggs and larvae include caddisfly and dragonfly nymphs, diving beetles, and red-spotted newts. Adults are probably preyed upon by shrews, raccoons, garter snakes, ringneck snakes, and ground-foraging birds. Blue-spotted salamanders have elaborate antipredator defensive behaviors, which are seen especially in juveniles and small adults. The salamanders raise their tails vertically, spread and brace their rear legs, and raise their vents off the ground. They will also wag their tails and secrete a sticky white substance from the tail.

Conservation Blue-spotted salamanders are common in many parts of their geographic range. Like many members of the genus *Ambystoma*, this species is vulnerable to changes in breeding pools, ponds, and other isolated wetland habitats.

Taxonomy The blue-spotted salamander, described in 1967, is not in question as a valid species. However, confusion concerning this species comes from its association with the unisexual *Ambystoma* complex, in which it is one of the major genetic contributors (see account on unisexual mole salamanders).

MOLE SALAMANDER *Ambystoma talpoideum*

BODY SHAPE
Chunky with a big
head; large legs;
short tail

**NUMBER
OF COSTAL
GROOVES**
10 or 11

BODY COLOR
Brown, gray, or
black

PATTERN
None, or small gray
flecks on the back
and tail

BODY SIZE
Metamorphs 2.2
inches; adults up to
4.5 inches in total
length

Description Adults and juveniles are chunky with a large, wide head and large feet; the tail is short for their size. They are light brown, various shades of gray, or nearly black with light gray specks on the back, sides, and tail. Some individuals lack the speckling altogether. The belly of older individuals is uniformly light gray to cream; recently metamorphosed individuals have a belly stripe.

Larvae and Juveniles Larvae are gray to olive brown with two creamy to yellowish stripes on the body that fade to blotches on the tail. The upper stripe may or may not be conspicuous, and the lower one may fade into the belly but is usually distinct. A narrow dark gray to black line, often bordered by parallel pale yellowish lines, is usually present along the middle of the belly. Juveniles

top An adult mole salamander showing the chunky head and body and typical color pattern

top Mole salamander larvae are gray to olive brown with two cream to yellowish stripes on the body that fade into blotches on the tail. The belly has a dark stripe along the midline.

bottom Individual color variants resulting from genetic anomalies, such as this larval mole salamander lacking black pigment except in the eyes, are likely to occur in many populations but at low frequencies.

are often greenish brown and may retain the yellowish stripe on the sides but may also show some gray flecking.

Similar Species Adult marbled salamanders are also chunky but have white to cream crossbars on the back. Their juveniles are brown to black with tiny gray to silver specks that may occur as a lichen-like pattern or as paired faded, yellowish spots on the back. Spotted salamander juveniles are either olive to gray with fine gray specks on the back but not on the tail, or they have the spots typical of adults. Immature small-mouthed salamanders are brownish gray to black with clusters of light specks or a lichen-like pattern mostly concentrated on the sides. Newly metamorphosed juvenile streamside salamanders are brown to brownish gray with little or no pattern.

The silver flecking on this immature mole salamander will largely disappear as it matures.

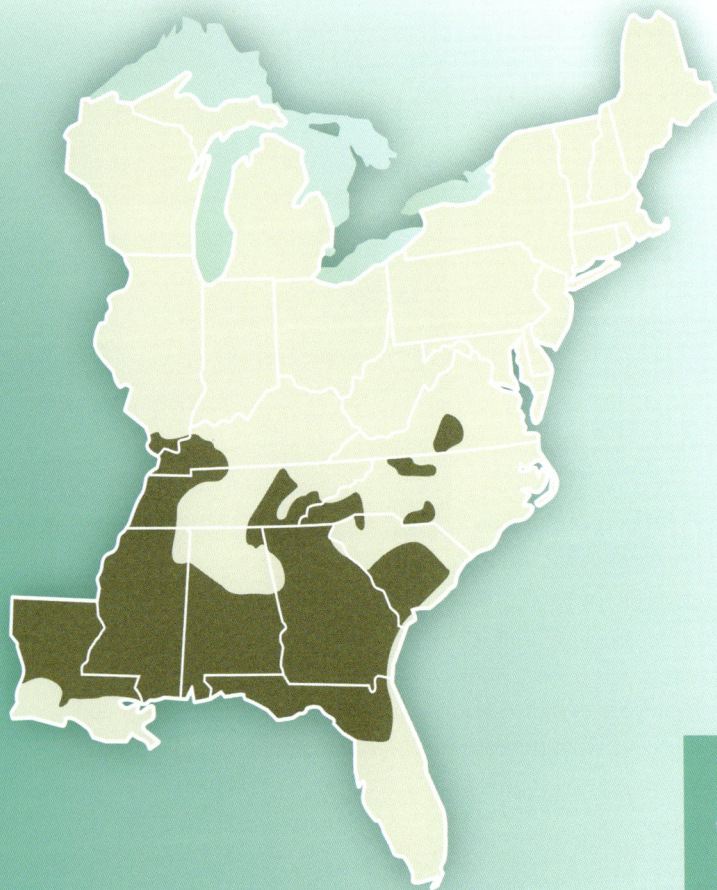

Distribution Disjunct, isolated populations of mole salamanders occur from south-central Virginia southward to northern, southern, and western North Carolina; in central to southern South Carolina; throughout Georgia; and from north-central and peninsular Florida westward through southern Alabama and all of Mississippi. Additional scattered, isolated populations occur throughout Louisiana, eastern Texas, Oklahoma, and Arkansas. The range of the mole salamander also extends up the Mississippi River drainage from eastern Louisiana through Mississippi to western Tennessee, Kentucky, and extreme southern Illinois.

The abundant silver flecking on the tail and sides of this adult mole salamander illustrate individual variation in pattern.

Habitat Mole salamanders occupy bottomland hardwood and floodplain forests near gum and cypress swamps and Carolina bays in the Coastal Plain. They also occur in hardwood forest and mixed hardwoods and pine forest around vernal pools, borrow pits, ditches, and other ponds outside the Coastal Plain. Breeding ponds usually lack fish and dry up in summer.

Behavior and Activity Adults remain underground in refugia or beneath pine straw or other ground litter when not in breeding ponds. Adults in the southern portion of the range migrate from the upland forests to breeding ponds from November to January during nighttime rains. Those in northern locations move to ponds during the late winter months. Mating occurs in the water immediately after arrival. Adults leave the ponds within several days to a week and migrate back to their forest refugia. Populations of neotenic individuals remain in the pond year-round. If the pond dries, they will absorb their gills and move into forest refugia.

Reproduction Terrestrial adults without gills mate and lay eggs in October through March, with the timing depending on weather conditions and the geographic area. Populations of neotenic gilled adults mate as early as September and as late as November. Some populations in permanent ponds have both neotenic and terrestrial breeding adults. Males deposit about 15 spermatophores on the substrate and entice females to pick up 1 or more of them with the cloaca. Females in Gulf Coast populations deposit 4–99 eggs in small clusters on twigs; those in Atlantic Coast states lay them singly on the pond bottom. Clutch size is 200–700 per female. Neotenic females are usually smaller than terrestrial females and lay fewer eggs. The gilled adult popula-

above Some individuals in southeastern populations remain in the water for long periods and acquire the ability to reproduce before metamorphosis. These paedomorphic mole salamanders will lose their gills and emerge from their pond when their habitat dries up.

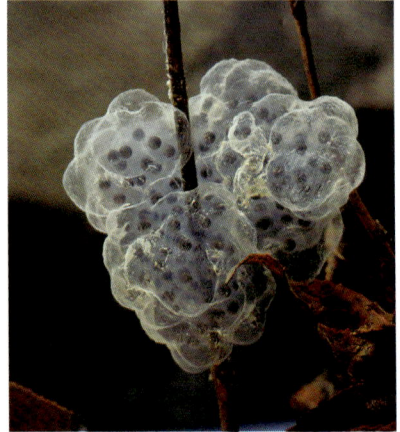

right Female mole salamanders in some areas attach a clutch of eggs to twigs and branches while females in other areas deposit single eggs on the substrate of the breeding pond.

tion will produce mature individuals by the following September. Larvae from adults without gills metamorphose in May to September but may not mature for several more months.

Food and Feeding Aquatic larvae consume a wide variety of small-bodied invertebrate prey, tadpoles, and other salamander larvae, including their own species. Adults probably eat earthworms, leeches, and other invertebrates.

Predators and Defense Fish, especially bluegills, are the primary predators of eggs and larvae, which hide in the leaf litter when fish are present. Watersnakes (southern banded, plain-bellied, and eastern green), black swamp snakes, cottonmouths, black racers, mud snakes, rainbow snakes, ribbon snakes, garter snakes, and wading birds eat adults. When threatened, adults use a head-down posture, become immobile, bite, head butt, flip the body, or wave the tail. Glands in the tail and the parotoid gland behind the head produce noxious secretions that may deter predators.

Conservation Mole salamanders are listed as a Species of Special Concern in North Carolina and as a Species in Need of Management in Tennessee. Primary threats include urban sprawl, agricultural practices, deforestation, and changes in landscape hydrology.

Taxonomy The mole salamander was first described in 1838. No subspecies are recognized.

SMALL-MOUTHED SALAMANDER *Ambystoma texanum*

BODY SHAPE	NUMBER OF COSTAL GROOVES	BODY COLOR	BODY SIZE
Chunky; small, short head and small mouth	13–15	Gray, brown, or black	Metamorphs 2.2 inches; adults up to 6 inches in total length
		PATTERN Pale, lichen-like	

Description The noticeably small head and mouth distinguish this species from other species of mole salamanders. Adults are shiny gray to grayish black with light gray, lichen-like speckling on the back and sides that is often concentrated on the sides. Some individuals are almost uniformly black. The belly is gray with light speckling or black.

Older larvae of small-mouthed salamanders are light brown with inconspicuous spots.

Larvae and Juveniles Hatchlings and small juveniles are brownish to olive green with three to six pairs of yellowish olive green patches along the middle of the back.

top This adult small-mouthed salamander has the species' characteristic small head and a uniformly gray body.

Older larvae are light brown with inconspicuous spots. Metamorphs are brown to gray and unmarked; the lichen pattern appears a few weeks after metamorphosis.

Similar Species Other mole salamanders have a normal-sized head. Marbled salamander juveniles have a lichen-like pattern of tiny gray to silver specks on a brown to black body or a series of paired faded yellowish spots on the back. Spotted salamander juveniles are olive to gray with fine gray specks on the back but not on the tail. Larval mole salamanders are gray with two creamy

right The pattern on the back of some small-mouthed sal-amanders is distinct from the pattern on the sides.

below Adult small-mouthed salamanders may be uniformly brown, although the gray phase is more common.

to yellowish stripes on the body that fade to blotches on the tail. Newly meta-morphosed juvenile streamside salamanders are brown to brownish gray with little to no pattern.

Distribution Small-mouthed salamanders occur in western Alabama and Ten-nessee and in the states bordering the Mississippi River southward to the Gulf coast of Louisiana and eastern Texas. They range west to eastern Texas, Okla-homa, Kansas, and extreme eastern Nebraska, and north to southern Iowa, southern Illinois, all of Indiana, and most of Ohio.

Habitat Adults and juveniles occupy bottomland hardwood forests along rivers and streams, and hardwood forests around ephemeral ponds and swamps. Juveniles hide under leaf litter and objects such as logs and debris.

Behavior and Activity Adults live and forage beneath the forest floor. They emerge en masse on rainy nights in late winter to early spring and migrate to breeding ponds. Reproduction occurs in a wide variety of shallow, temporary wetlands such as woodland pools, ditches, borrow pits, flooded fields, and swamps. Juveniles spend most of their time under the leaf litter.

Reproduction Males usually arrive at breeding ponds before females during winter or early spring and place spermatophores on leaves, twigs, and other bottom features. Females arrive shortly afterward. As each is courted by a male, she uses her cloaca to pick up a sperm packet sitting atop a spermatophore; fertilization is internal. One female may pick up several sperm packets, not all necessarily from the same male. Females deposit 250–800 eggs singly, in loose clusters, or in gelatinous clusters of 4–30 eggs each. Hatching occurs in later winter. The larvae remain in the pool for 2–4 months, then emerge during rain as metamorphs.

Small-mouthed salamander females lay their eggs in small clusters attached to twigs and stems in the water.

Food and Feeding Adults and juveniles eat a wide variety of prey including earthworms, centipedes, moths, crickets, and beetles. Larvae eat small invertebrates such as isopods and copepods and may be cannibalistic. The size of the salamander's mouth dictates the size of the prey it can swallow.

Predators and Defense Known predators of larvae include tiger salamanders, predaceous insects, garter snakes, and watersnakes. When threatened, adults lower the head, curl the body, and wave the tail in an attempt to disperse noxious gland secretions that may discourage a predator. Larvae also use concealment and rapid swimming to escape.

Adults are shiny gray to grayish black with light gray, lichen-like speckling on the back and sides that is often concentrated on the sides.

Conservation Urbanization and agriculture threaten wetland breeding sites of small-mouthed salamanders and the adjacent forest cover where they spend most of their lives. Many adults and metamorphs are killed on roads during migrations to and from breeding ponds. No southeastern state wildlife department classifies this species as Endangered or Threatened.

Taxonomy The small-mouthed salamander was described initially in 1855. No subspecies are recognized.

TIGER SALAMANDER *Ambystoma tigrinum*

BODY SHAPE
Robust; large head;
long tail

**NUMBER
OF COSTAL
GROOVES**
11–14

BODY COLOR
Olive gray to black
or brown

PATTERN
Yellowish irregular
spots on body, head,
and tail

BODY SIZE
Metamorphs 2–4
inches; adults up
to 13 inches in total
length

Description Adult tiger salamanders are the largest mole salamanders in the eastern United States—up to 13 inches in total length. The body is robust, and the head is large. The tail of males is compressed side to side and is larger than that of females. The body is bluish gray to brown to nearly black. The back, sides, and tail have olive to yellowish brown, irregularly shaped spots. The belly is olive yellow to cream with some faded dark pigment in a marbled pattern.

Larvae and Juveniles Larvae are olive to gray with paired dark blotches on the body and tail and a white belly. Their toes are wedge-shaped; those of all other mole salamander larvae are round. Juveniles are olive gray to dark brown with at least some irregular spots on the body.

top Tiger salamanders are the largest terrestrial salamanders in the eastern United States.

above Three tiger salamander juveniles illustrate some of the range of variation in body pattern at this early stage.

left Larval tiger salamanders are olive to gray with pairs of dark blotches on the body and tail. Each toe is wedge shaped.

Similar Species Marbled salamander juveniles are brown to black with tiny gray to silver specks that may occur as a lichen-like pattern or as pairs of faded yellowish spots on the back. Spotted salamander juveniles are olive to gray with fine gray specks on the back but not on the tail. Juvenile Jefferson salamanders are uniformly dark gray, occasionally with brownish yellow specks on the sides. Immature small-mouthed salamanders are brownish gray to black with clusters of light specks or a lichen-like pattern mostly concentrated on the sides. Newly metamorphosed juvenile streamside salamanders are brown to brownish gray with little or no pattern.

Distribution Tiger salamanders are found along much of the East Coast from Long Island, New York, to southern New Jersey; in scattered, isolated localities in Virginia; and through the Coastal Plain of North Carolina, South Carolina, and Georgia southward to north-central Florida. They occur in isolated locations throughout the Southeast. In the Midwest they have a much wider distribution that includes northwestern Ohio, southern Michigan, and all of Indiana, Illinois, Iowa, and Minnesota. They also occur in western Wisconsin.

Habitat Adults live almost anywhere there is forest cover and soil in which they can burrow; they primarily occupy hardwood forest habitats but may also occur in longleaf pine forests, mixed pine and hardwood forests, and bottomland forests. Breeding takes place in seasonal to permanent ponds that lack fish, as well as in roadside ditches, artificially created wildlife ponds, and slow-moving streams.

TIGER SALAMANDER *Ambystoma tigrinum*

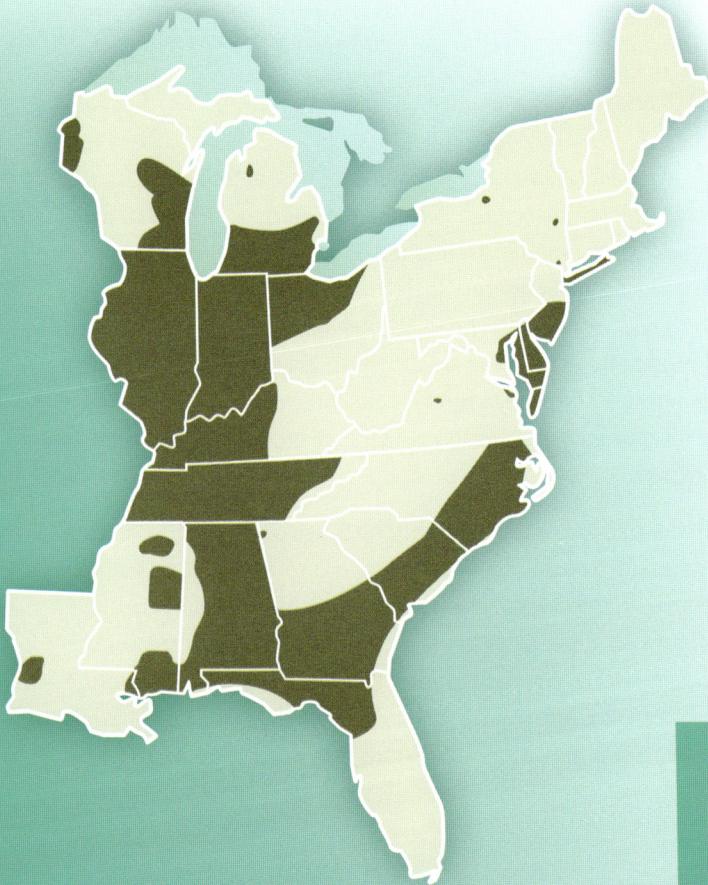

The spotted pattern of this adult tiger salamander is found in some eastern populations.

Behavior and Activity Unlike some other mole salamanders, tiger salamanders dig their own burrows in the soil. They remain underground—and are seldom seen—between breeding seasons and migrate to breeding ponds during warm rains in winter months. Breeding ponds are usually less than 300 yards from their upland refugia, although such distances vary depending on habitat type and other aspects of the landscape.

Reproduction Males arrive as early as late December and mate in the water with females when they arrive 2–4 weeks later. Mating is sometimes under ice on the pond's surface. The males deposit spermatophores on the pond substrate that females pick up with the cloaca and use to fertilize their eggs. Gelatinous masses of 5–122 eggs are attached to twigs. Adults leave the pond after mating and return to their burrows. The larvae hatch several weeks after the

Developing tiger salamander eggs are visible within the egg capsule and jelly-like egg mass.

above Tiger salamanders will eat any animal they can capture and subdue. Their big head allows them to take large prey.

left The broad head and upwardly curved mouth make this salamander appear to be smiling. Note the nostrils and the widely spaced eyes used for visual detection of prey.

eggs are laid and reach metamorph size in 10 weeks to 6 months.

Food and Feeding The larvae consume zooplankton, a wide variety of aquatic invertebrates, worms, leeches, crayfish, tadpoles, other salamander larvae, and sometimes each other. Individuals in some populations become larger than the other tiger salamanders, grow larger teeth, and become cannibals. Adults do not feed in the breeding season but in their upland habitats eat any prey they can get in their mouth, including a wide variety of invertebrates and even mice.

Predators and Defense Known aquatic predators include predaceous diving beetles, dragonfly nymphs, other tiger salamanders, southern banded water-snakes, mud snakes, and wading birds. Terrestrial predators include shrews, middle-sized mammals such as raccoons, and snakes. Larvae hide during the day in vegetation and under leaf litter at the bottom of the pool and forage in the open water column on moonless nights. Adults produce a sticky, toxic secretion in glands in their tail that repels and may even kill predators.

Conservation Eastern tiger salamanders are protected as Endangered in Alabama, Delaware, Maryland, New Jersey, New York, and Virginia and as Threatened in North Carolina. The primary threat is loss of both upland habitat and pond breeding sites to urban sprawl. Populations at high elevations are affected by acid precipitation. Mortality can be extremely high on roads between breeding sites and upland habitats. Several diseases are known to occur in some populations, but their impacts are unknown. Upland and breeding habitats in areas where populations are in danger of extirpation need proper management if the tiger salamanders are to be preserved.

Taxonomy   The tiger salamander was first described in 1825. The several subspecies recognized in western North America have been grouped within the barred tiger salamander species (*Ambystoma mavortium*) complex, leaving the eastern tiger salamander without subspecies.

Adult tiger salamanders produce a sticky, toxic secretion in glands in their tail that repels and may even kill predators.

MABEE'S SALAMANDER *Ambystoma mabeei*

BODY SHAPE
Slender; small head

NUMBER OF COSTAL GROOVES
13

BODY COLOR
Black to dark brown

PATTERN
Round patches to scattered patches of gray flecks

BODY SIZE
Metamorphs 2.2 inches; adults up to 4.5 inches in total length

Description Adults are medium-sized mole salamanders with a small head and small round patches of light gray to silver specks all over the black body, head, and tail. The markings run together and concentrate along the sides of the body and are relatively sparse on the back, but the individual pattern is highly variable. The belly is dark brown to gray. As in other species of mole salamanders, males in breeding condition have swollen cloacal glands; the cloacal area of females is flat.

Larvae and Juveniles The aquatic larvae have bushy gills, dorsal fins that extend almost to the front limbs, and a yellowish stripe bordered above and below with black along each side. Older larvae have a second yellow stripe below the lower black zone, and both stripes may be broken or appear blotched. The belly

top Adult Mabee's salamanders have a uniformly gray back with abundant silvery flecking and markings on the sides.

is cream-colored, and the tail fin has dark black mottling. The specks on juveniles are small and not clustered as in older adults, and are not concentrated along the sides of the body. Larvae transform into terrestrial juveniles after 2–9 months, depending on when their woodland ponds fill with water and subsequently dry.

Mabee's salamander larvae are gray to olive brown with two cream to yellowish stripes on the body that fade into blotches on the tail.

Similar Species Mole salamanders are gray and chunky with a large head. Spotted salamanders are black with yellow spots, and the juveniles are gray with fine black specks on the back but not on the tail. Juvenile marbled salamanders are brown to black with tiny gray to silver specks that change into a lichen-like pattern within a month after metamorphosis.

Distribution Mabee's salamanders occur in the Atlantic Coastal Plain from southeastern Virginia to the southern portion of the Savannah River in South Carolina. No records are known in the Piedmont, north of the Fall Line.

Habitat Adults and juveniles occupy animal burrows and root tunnels in open fields, pine forests, and hardwood forests when not at the breeding pond. They may be found under all manner of surface cover objects during wet periods. Breeding sites include vernal pools, ditches, small fishless ponds, and Carolina bays.

Behavior and Activity Adults and juveniles may be active on the surface during rains but otherwise remain underground. Surface activity increases as the breeding season approaches. In January, February, and March they migrate

Mabee's salamander metamorphs are uniformly gray to brown and develop silvery flecking.

from their terrestrial refugia to wetlands up to half a mile away where they mate and lay eggs. Adults remain in the pond for several days to weeks after breeding and then migrate back to their upland refugia, usually on a rainy night.

Reproduction The courtship behavior of this salamander has not been published. Females attach eggs singly or in strings of 2–6 to leaves and other material on the bottom of the pond. Individual female clutch size is unknown.

Food and Feeding The larvae feed on zooplankton and small invertebrates. Mouth size determines the size and type of prey that they can eat. Adults are known to eat earthworms in upland habitats.

Predators and Defense The larvae are eaten by tiger salamander larvae, lesser sirens, fish, dragonfly nymphs, and predaceous water beetle larvae. Predators

MABEE'S SALAMANDER *Ambystoma mabeei*

This subadult Mabee's salamander has profuse silver flecking and spotting on the back that will fade as it ages.

of adults have not been observed but likely include shrews and snakes that find them underground. When threatened, adults become immobile, hide the head behind the tail, curl and wave the tail to distract the predator, and tuck the head down to the substrate to minimize exposure.

Conservation Mabee's salamanders are listed as a Threatened species by the states of North Carolina and Virginia. Threats include habitat loss, road mortality, and alteration of pond hydrology resulting from ditching for urban developments.

Taxonomy Mabee's salamander was first described in 1928. No subspecies are recognized.

MARBLED SALAMANDER *Ambystoma opacum*

BODY SHAPE
Chunky

**NUMBER
OF COSTAL
GROOVES**
11–13

BODY COLOR
Black

PATTERN
White or light gray
crossbands across
the head, back, and
tail

BODY SIZE
Metamorphs 1.8–2
inches; adults up to
4.5 inches in total
length

Description The fully terrestrial adults are medium-sized and have a chunky
build. The black ground color has an overlay of contrasting white or light gray
crossbands across the head, back, and tail. The crossbands may appear as
stripes on some individuals, and rare individuals are entirely black. The belly is
uniformly black. The crossbands are silvery white on males and silvery gray on
females. Males in breeding condition also can be distinguished from females
by their swollen cloacal glands.

Larvae and Juveniles The aquatic larvae have bushy gills, a dorsal fin that
extends almost to the front limbs, and are drab brown or black with a series
of light spots that form a line on the sides. Larvae transform into terrestrial
juveniles after 2–9 months, the period depending on geographic location and

top This adult male marbled salamander has bright crossbands on his body and tail and abundant white pigment on his head. Individual salamanders can be recognized by their pattern.

bottom Metamorphic marbled salamanders develop a highly variable amount of frosty flecking that becomes crossbars and other patterns in adults.

when their woodland ponds fill with water and subsequently dry.

Similar Species Spotted salamander juveniles are dull olive green to dark gray with fine gray specks on the back but not on the tail. Mole salamanders are short and chunky with a large head and light gray or no flecking. Juvenile Jefferson salamanders are uniformly dark gray, occasionally with brownish yellow specks on the sides. Streamside salamander juveniles are uniformly brown to brownish gray.

Distribution Marbled salamanders occur in much of the eastern United States from eastern Texas and Oklahoma northeast through Illinois and Indiana to southern New Hampshire and central Massachusetts, and south to northern Florida.

Habitat Mating takes place at night in the open on the surface, and females lay eggs underneath cover objects in the dry basins of temporary, fish-free wetlands such as upland hardwood "swamp forests," bottomland hardwood pools, quarries, vernal ponds, Carolina bays, and floodplain pools. Outside the autumn breeding season terrestrial adults and juveniles live underneath leaf litter and in small mammal burrows in mature deciduous forest, mixed hardwoods, pine stands, floodplains, and uplands.

Behavior and Activity Marbled salamanders may be inactive for prolonged periods in the summer when the weather is hot and dry. On rainy nights from September through November adults migrate to the dry basins of temporary wetlands; males typically arrive at breeding sites before females. The adults return to the wooded habitat surrounding the wetland on rainy nights between late fall and winter.

Females lay their eggs in depressions around the margin of a breeding pond under leaves, moss, or logs and remain with them until the pool fills with water.

Reproduction Marbled salamanders are one of only two species of mole salamanders that breed on land, and they are the only member of the genus *Ambystoma* in which the female parent guards the developing embryos. Following courtship, females lay from 30 to nearly 200 single eggs in nests constructed underneath logs, vegetation, or leaf litter in the dried beds of ponds, the margin of reduced ponds, or dry floodplain pools, typically in a spot likely to be flooded by winter rains. Most females nest singly, although two to seven females have been observed nesting communally under the same cover object. Females typically stay with their nests to guard the eggs until the pond begins to fill and the larvae start to hatch; however, nests without attending females are not unusual.

Food and Feeding Adults feed on terrestrial invertebrates such as millipedes, centipedes, spiders, insects, snails, arthropods, worms, and mollusks. The larvae eat small aquatic invertebrates—including a variety of crustaceans such as copepods, amphipods, and isopods—and smaller salamander larvae.

Predators and Defense The most common known predators of adults are eastern garter snakes, short-tailed shrews, raccoons, and opossums. The larvae are eaten by southern banded watersnakes, eastern green watersnakes, and black swamp snakes. Adults have granular glands on the head that produce noxious secretions known to deter attacks by shrews and common ribbon snakes.

Marbled salamanders show considerable variation. This individual has two stripes down its back rather than the crossband pattern typical of most adults.

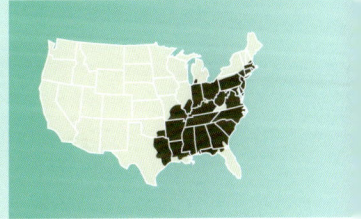

Conservation Loss of bottomland hardwoods and their associated wetland habitats poses the greatest risk to this species. Small, isolated wetlands are the most valuable habitats for maintaining high amphibian biodiversity, but they are also most at risk because current regulations generally do not protect them. Marbled salamanders are listed as Endangered in Michigan and as Threatened in Massachusetts.

Taxonomy The marbled salamander was first described in 1807. No subspecies are recognized.

SPOTTED SALAMANDER *Ambystoma maculatum*

BODY SHAPE
Robust body;
rounded head

**NUMBER
OF COSTAL
GROOVES**
11–13

BODY COLOR
Black

PATTERN
Distinct yellow
spots

BODY SIZE
Metamorphs 2
inches; adults up
to 9 inches in
total length

Description Spotted salamanders are large and robust with a black body and distinct yellow or orange spots on the body, head, and tail. The belly is uniformly dark gray.

Larvae and Juveniles The larvae are greenish to gray above with few markings and have an unmarked white belly. At metamorphosis, the robust juveniles have a thin-lined, reticulate pattern on a gray background or are uniformly gray with no markings. The spots appear within several months.

Similar Species Larval marbled salamanders have scattered specks of dark pigmentation on the throat. Juvenile marbled salamanders are brown to black with tiny gray to silver specks that may occur as a lichen-like pattern on the

top Adult spotted salamanders typically have yellow spotting on the head, body, and tail on a dark gray body.

back. Small-mouthed salamander adults are brownish gray to black with clusters of light specks or a lichen-like pattern mostly concentrated on the sides. Mole salamanders are short and chunky with a large head and are nearly uniform in color except for a scattered pattern of light gray flecks. Juvenile Jefferson salamanders are uniformly dark gray, occasionally with brownish yellow specks on the sides.

Distribution Spotted salamanders occur widely in the eastern United States from Canada south through Virginia and the Carolinas to the Gulf of Mexico in Alabama and Mississippi, and westward to eastern Texas and southeastern Missouri. They are absent from the Coastal Plain of North Carolina, southeastern Virginia, southern Georgia, and most of Florida.

DID YOU KNOW?

Salamander larvae are voracious predators of mosquitoes and serve as natural biological control agents in pools and other water bodies. They make far better mosquito control agents than introduced mosquito fish, which eat salamander larvae and eggs.

above The yellow spots are beginning to appear on this juvenile spotted salamander, which is several months old.

left top Typical spotted salamander larvae are brown to gray with bushy gills. They differ from marbled salamander larvae in lacking black pigment on the throat.

left bottom Metamorphic spotted salamanders are uniformly gray with few markings. The dark remnant of the gill bud on the back of the head indicates that this little salamander is a recent metamorph.

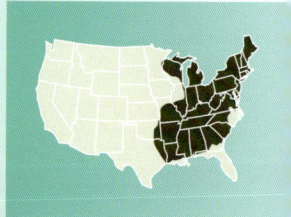

This is a rather unusual unspotted spotted salamander from northern Georgia.

above and right Adult spotted salamanders in some populations have orange spots, often on the head, in addition to the yellow spots on the body and tail.

Habitat Spotted salamanders live primarily in hardwood and mixed deciduous forest habitats but may be found in a wide array of habitat types, including mesic floodplains, if forest cover is present. They breed in shallow fresh water, often in isolated and ephemeral wetlands. The combination of wetlands and upland terrestrial habitats is essential to the survival of this salamander.

Behavior and Activity Late-winter rains bring adults from forest refugia to shallow breeding ponds. Eggs hatch in 1–3 weeks, and the larval period lasts about 3 months depending on how elevation, latitude, and canopy cover affect water temperatures. Post-metamorphic animals migrate away from wetlands in mid-to-late summer. Adults and juveniles hibernate in subterranean burrows in the forest and may move down below the frost line in winter to avoid freezing. They may be found under logs and other objects on the forest floor, especially after rains.

Female spotted salamanders lay egg masses that may be either clear or opaque. The difference is based on one gene difference among members of the population.

Reproduction Males arrive first at the breeding pond and deposit several spermatophores (sperm packets) on the substrate. Females arrive shortly afterward and are courted by the males, which use touch and chemical cues to position females over one of their spermatophore packets. The eggs are fertilized as they pass through the female's cloaca while

True albino spotted sala-
manders with red eyes and
no black pigment are rare
but may occur anywhere in
the species' range.

being deposited. The female attaches one to several gelatinous egg masses containing up to 250 eggs to a stem or twig in shallow water. The egg jelly is clear
or opaque white. Egg masses that appear green have a symbiotic alga inside
individual eggs that may increase the oxygen supply to developing embryos.

Food and Feeding The aquatic larvae eat small invertebrates such as cladocerans, zooplankton, beetles, isopods, ostracods, dragonfly and damselfly larvae,
and caddisflies; and sometimes each other. Adults consume a wide variety of
invertebrates—such as worms, centipedes, millipedes, spiders, and numerous
insects—and the occasional small mouse.

Predators and Defense Various fishes, red-spotted newts, wood frog (*Rana
sylvatica*) tadpoles, and aquatic larval and adult invertebrates—including caddisflies, midges, predaceous diving beetles, giant water bugs, water tigers, and
dragonfly nymphs—prey on eggs and larvae. Marbled salamander larvae, which
are larger, often occur in the same wetlands and will eat spotted salamander
larvae. Wood frog tadpoles, snakes, wading birds, raccoons, and possibly opossums, skunks, and foxes eat juveniles and adults. Egg masses are coated by a
thick jelly that may deter some predators; larvae and adults use concealment
and noxious skin secretions for defense.

Conservation Spotted salamanders are declining in many areas as a result
of deforestation and loss of wetlands, acid precipitation, heavy metal concen-
trations, environmental contaminants, and fish introduced into the wetlands
where they breed. Acid precipitation in the Appalachian Mountains threatens
many populations. Effective ways to mitigate elevated acid levels in breeding
ponds need to be developed.

Taxonomy The spotted salamander was first described in 1802. No subspecies
are recognized.

Description Unisexual mole salamanders are an interesting assemblage of primarily female salamanders with genomes derived from two or three of at least five different species in the genus *Ambystoma*: the streamside salamander (*Ambystoma barbouri*), the Jefferson salamander (*Ambystoma jeffersonianum*), the blue-spotted salamander (*Ambystoma laterale*), the small-mouthed salamander (*Ambystoma texanum*), and the tiger sal-

amander (*Ambystoma tigrinum*). These salamanders are sometimes referred to as unisexual hybrids, although the term "hybrid" technically should only be used for the first set of offspring produced from the mating of two distinct species. These salamanders, in contrast, are of multiple derivation, and aside from some extremely rare and short-lived males, they are all females. They cannot breed without the presence of males from bisexual spe-

A typical unisexual *Ambystoma* from northwestern, Ohio

top A unisexual *Ambystoma* with blue-spotted salamander and tiger salamander DNA from Kelley's Island, Ohio

cies and therefore defy conventional definitions of what constitutes a species. Most unisexual mole salamanders are diploid (having two complete sets of homologous chromosomes) or triploid (having three sets of homologous chromosomes), but tetraploid and even pentaploid individuals have been found. Adults are usually intermediate in appearance between their parents and populations can usually be identified as unisexual, but their species combination can only be determined through genetic analysis.

Unisexual mole salamanders do not represent a species or group of species because they depend on the males of other, bisexual *Ambystoma* species for reproduction. They are also not a group of populations consisting of distinct male and female cohorts between which there is gene exchange (that is, they do not breed with other unisexual mole salamanders), nor are they morphologically distinct from other *Ambystoma* species. As far as is known, they do not reproduce by parthenogenesis, which would mean that female offspring were produced from unfertilized eggs.

The nuclear genome (all of the DNA in the cell nucleus) of a unisexual mole salamander may incorporate genomes from up to three different *Ambystoma* species. The unisexual genotypes are named according to which genomes they

left A triploid (LLJ) unisexual *Ambystoma* from Connecticut with blue-spotted salamander and Jefferson salamander DNA

below A unisexual *Ambystoma* from an island in Lake Erie, Ohio, with blue-spotted salamander and small-mouthed salamander DNA

possess and are usually abbreviated as B, J, L, T or Ti (representing *A. barbouri*, *A. jeffersonianum*, *A. laterale*, *A. texanum*, and *A. tigrinum*, respectively). The term "biotype" has been used to identify each unique combination of genomes in the unisexual *Ambystoma*. More than 20 different biotypes have been identified that have at least one set of chromosomes from *A. laterale*. It was once thought that *A. laterale* was the original sperm donor, but current thinking is that the *A. laterale* genome may instead have a regulatory role, silencing incompatible genes of other nuclear genomes. The spotted salamander (*A. maculatum*) occurs with many populations of unisexual *Ambystoma*. However, its genome has never been found in any unisexual salamander, perhaps indicating that it might be lethal to a unisexual genome.

Several hypotheses attempt to explain the evolutionary origin of unisexual mole salamanders. The most widely accepted and supported hypothesis is that unisexual mole salamanders represent a single lineage that split from other mole salamanders at least 2–5 million years ago. The unisexual mole salamanders persist as a genomic "parasites" and have been described as kleptogenic in their mode of reproduction. Kleptogens are females that maintain a common cytoplasm across the group but have nuclear genomes that can vary. Genomes are acquired from males that have a genome compatible with the kleptogen cytoplasm. Once acquired, the genetic contents of the sperm nucleus may or may not be incorporated into the offspring's nuclear genome, and if incorporated, the set of chromosomes from the sperm may replace one of the female's sets, or it may add an additional chromosome set, thereby increasing the level of ploidy (number of chromosome sets).

The various unisexual mole salamander biotypes arise by three different mechanisms: gynogenesis, wherein diploid eggs from a unisexual female are activated but not fertilized by sperm from the male of a bisexual species; hybridogenesis, in which the active genome of a unisexual female is fertilized by the sperm and results in the replacement of one of the maternal genomes; and finally, activation and fertilization of the diploid eggs by the sperm with the addition rather than replacement of a genome. Genome replacement through hybridogenesis is common among unisexual mole salamanders and probably serves to compensate for a lack of genetic diversity that can result from the absence of sexual reproduction.

Larvae and Juveniles Metamorphosis occurs from June to late August with different biotypes having different rates of development. Bisexual larvae tend to spend more time in shallower water within vegetation, and the unisexual larvae are often found in deeper, more open portions of ponds.

Similar Species Unisexual mole salamanders look very much like the bisexual species of *Ambystoma* with which they share habitat. Differentiating them from any of those five species (*A. laterale, A. jeffersonianum, A. barbouri, A. texanum,* or *A. tigrinum*) is challenging and often requires molecular analysis.

Genetic analysis of the nuclear genome is the most reliable technique to distinguish unisexual from bisexual *Ambystoma*. Polyploid unisexual salamanders (those having more than two sets of chromosomes) can be identified by counting the chromosomes in the red blood cells. Often in areas where unisexuals occur the sex ratio has more females than males in the population. Where unisexuals are present there are often many dead egg masses due to females laying eggs without male contact or mating with males that have incompatible genomes.

UNISEXUAL MOLE SALAMANDERS

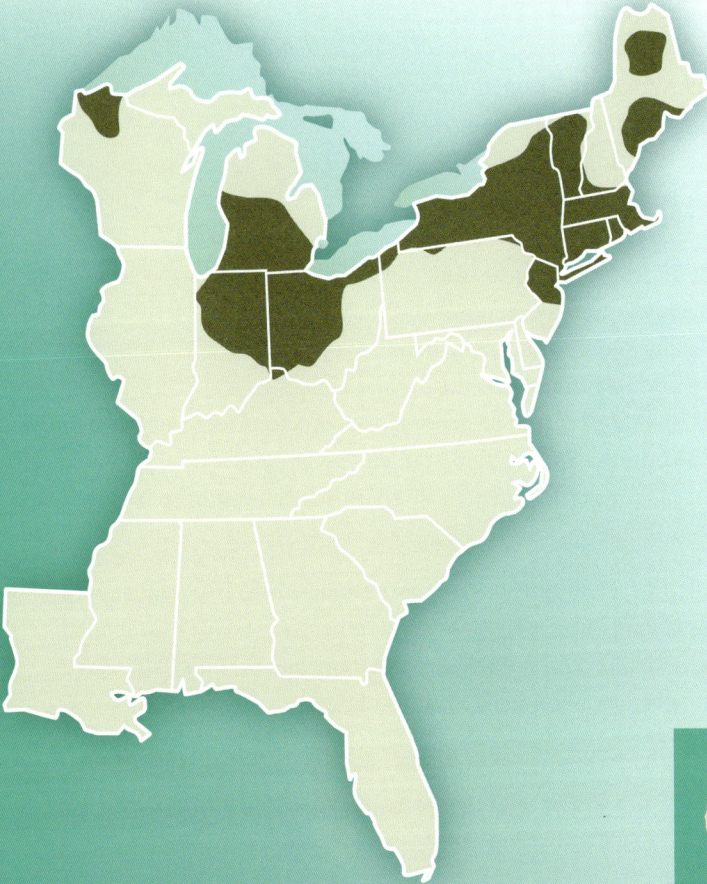

Distribution Unisexual mole salamander populations have been known
since 1934. They are concentrated in
the Great Lakes region, including
parts of Wisconsin, Michigan, Indiana, Ohio, Pennsylvania, New York,
southeastern Ontario, and southern
Quebec, but also occur in parts of New
Jersey, the New England states, and
southeastern Canada on the Atlantic
Coast. LLJ unisexuals are the most
widely distributed biotype. LJJ unisex-

A unisexual (LLJ) *Ambystoma* of hybrid derivation from Michigan

uals are found only in Indiana, southern Michigan, Ohio, extreme northern
Kentucky (in Kenton, Boone, and Pendleton Counties), and southern Ontario.
Unisexual mole salamander individuals are often present where sympatric bi-
sexual *Ambystoma* species occur.

Habitat The unisexual mole salamanders are found in the same habitats
as their bisexual relatives, often under the same logs and rocks. Many of the
unisexual animals are found year-round near the surface under logs, rocks,
and leaves. Unlike salamanders of the genus *Ambystoma* that occur farther
south, they rarely dig and tunnel deep into the soil to avoid the summer heat
and drier climate.

Behavior and Activity The natural history of unisexual mole salamanders
mostly mirrors that of their bisexual counterparts.

Reproduction Unisexual mole salamanders, like their bisexual counterparts,
use woodland vernal ponds for their breeding sites. These temporary bodies of
water need to be free of fish and persist long enough for the larvae to metamorphose. Breeding occurs from mid-February to mid-April with some variation
along a north to south gradient. Typical breeding temperatures range from
39°F to 68°F with rain being the triggering factor of the breeding run. Since
unisexuals are all females, they migrate into the ponds when the males of bisexual species are arriving or have arrived. These males appear to recognize
unisexual females and do not spend as much time with, or do not actively court,
them. The unisexual salamanders locate spermatophores (mucus covered
packages of sperm deposited by the males) and pick them up for fertilization
or egg activation. Females straddle the spermatophore, picking it up with their
cloacal lips and bringing it into their oviducts. They can store the sperm for

A tetraploid (LLJJ) unisexual *Ambystoma* from New York

several days before using it to internally fertilize or activate their eggs. Eggs are then laid in the breeding pond, where they will develop externally without parental care.

Food and Feeding The diet of unisexual mole salamanders is thought to be the same as the bisexual mole salamanders that share their habitat. Earthworms and a variety of invertebrates comprise most of their prey items.

Predators and Defense Skunks, opossums, raccoons, and shrews are known predators of the adults, but snakes and owls are also potential predators. Adult unisexual mole salamanders elevate their tail, coil the body, hide their head under the tail, and produce a noxious secretion from tail glands that deter and confuse potential predators. Larvae are eaten by newts, watersnakes, wading birds, and aquatic predaceous insects.

Conservation The unisexual mole salamanders present a unique problem for conservation. Since they are not technically classified as "species" they are often overlooked on protected lists and are not considered in land management plans. They suffer the same conservation issues as do other ambystomatid salamanders, such as wetland drainage and alteration, disruption of migration routes from breeding to nonbreeding habitats, and urbanization. The only state that affords any protection status to the unisexual mole salamanders is Massachusetts. They list the unisexual salamanders of the genus *Ambystoma* as being of Special Concern and the bisexual species *A. laterale* (blue-spotted salamander) as Threatened.

Taxonomy Information and research are reshaping our understanding of this unique group of salamanders that continues to fascinate and confuse herpetologists. Our knowledge on this complex genetic phenomenon continues to advance as researchers refine molecular techniques.

FOUR-TOED SALAMANDER *Hemidactylium scutatum*

BODY SHAPE
Stout; constriction around base of tail

NUMBER OF COSTAL GROOVES
13 or 14

BODY COLOR
Reddish brown

PATTERN
Irregular black markings on back; black-and-white belly

BODY SIZE
Adults up to 4 inches in total length

Description Four-toed salamanders derive their common name from the four toes on each of the rear feet; most other salamanders have five toes. Adults are usually reddish to rusty brown on the head, body, and tail, and have small, irregular black spots. The sides have gray pigment. The cream to white belly with black spots is unique to this species, and the noticeable constriction at the base of the fragile tail is also a key character. Females usually have rounded snouts and their snout-to-vent length is 15 percent longer than the more square-snouted males.

Four-toed salamanders have four toes on each of the rear feet; most other salamanders have five.

top Four-toed salamanders are usually reddish above with small, irregular black spots.

The slender, aquatic larvae of four-toed salamanders are yellowish brown with prominent eyes.

Larvae and Juveniles The slender larvae are yellowish brown with prominent eyes and a tail fin that extends almost to the head. After metamorphosis, juveniles take on the color and pattern of the adults.

Similar Species Dwarf and two-lined salamanders are yellow with a dark stripe on each side and a plain belly. Red and mud salamanders have many or few black spots, respectively, on a red or rusty body, and an unmarked belly.

Distribution Four-toed salamanders have been found in every eastern state, but their range is highly fragmented geographically. The species ranges from Nova Scotia, Canada, westward to Minnesota and southward through the midwestern and southeastern states as far south as the Florida panhandle. It occurs along the Atlantic Coast from Nova Scotia through Virginia, but does not occur in most of coastal North Carolina or in coastal South Carolina and Georgia. It occurs only in small, isolated populations in the southern and midwestern states.

Habitat Adults and juveniles prefer mature hardwood forests, but occasionally pine forests, surrounding swamps, freshwater marshes, isolated woodland pools, and other shallow wetlands without fish. They often remain hidden under rocks, logs, and other surface objects, and under moss on logs. Sphagnum

left The constriction at the base of the tail of this four-toed salamander is characteristic of the species.

opposite Juvenile four-toed salamanders take on the color and pattern of the adult after metamorphosis.

moss is often associated with their breeding ponds but they can also use ponds with other moss species.

Behavior and Activity Adults remain hidden in the forest floor outside the egg-laying season but are occasionally found under logs and other surface objects. Females move to nesting ponds from the last week of March through the second week of April but depending on location and rain events can wait as late as early June. They return to their terrestrial underground retreats after their eggs hatch. Adults move on rainy nights in spring and summer and are occasionally found on roads. As many as two hundred individuals have been found at one site, under leaf litter and inside rotting logs, in November. It has been speculated that they spend the coldest winter periods underground.

Reproduction Mating occurs on land in late summer, fall, and early winter, but females do not lay eggs until they migrate to wetlands in late winter and early spring. The courtship ritual is very elaborate and can last up to twenty minutes. Each female lays 4–80 eggs in a crude cavity in a clump of moss at the waterline, under loose bark, or in a clump of grasses, attaching single eggs to rootlets and moss

top Female four-toed salamanders often remain with the eggs until they hatch and drop into the water below the sphagnum moss.

bottom Sphagnum moss is often associated with the breeding ponds of four-toed salamanders.

strands to form loose clusters. Several females sometimes lay their eggs together in one moss clump, and one or two of the females will remain with the eggs until they hatch and drop into the water below. Reproduction occurs earlier in southern populations than at higher latitudes and elevations. The larval period lasts 23–39 days, which is very short compared to most other salamanders, and metamorphosis occurs in late spring when the larvae are about an inch long.

Food and Feeding Larvae probably eat small aquatic invertebrates, but little is known about this stage of the life cycle. Juveniles and adults prey on small beetles, moths, spiders, mites, springtails, and worms captured in the forest leaf litter.

Predators and Defense Larval four-toed salamanders fall victim to predatory insects and the larvae of other salamanders such as marbled, spotted, and Jefferson salamanders, and newts. They do not have noxious skin secretions and are readily eaten by fish, a primary reason why adults avoid breeding sites with fish. Adults are eaten by shrews, skunks, small birds, and ringneck snakes. Defensive behaviors include coiling the body tightly and hiding the head, immobility, emitting noxious secretions from glands in the tail, and breaking the tail completely off at the constriction point. The severed tail distracts the potential predator and allows the rest of the salamander to escape. A new tail is regenerated soon afterward. The four-toed salamander is one of the few salamanders that can voluntarily lose its tail (tail autonomy) to avoid predation.

Conservation The four-toed salamander is thought to be in decline throughout its range due to its specialized habitat requirements. Loss of wetland

Defensive behaviors of the four-toed salamander include coiling the body tightly and breaking the tail completely off at the constriction point. The severed tail distracts the potential predator, allowing the salamander to escape.

The cream to white belly with black spots is a signature characteristic of the four-toed salamander.

breeding sites and hardwood forest surrounding them is the main threat to four-toed salamanders. Commercial moss collecting is an underappreciated threat. The low dispersal ability of this small salamander often prevents it from recolonizing suitable habitat once it has been extirpated from a habitat. This species is currently listed as Endangered in Minnesota, Threatened in Illinois, and as Special Concern or Rare in Indiana, Wisconsin, Ohio, and Missouri. It is not protected in any state in the Southeast except by regulations governing its collection.

Taxonomy The four-toed salamander was first described in 1838. No subspecies are recognized.

EASTERN NEWT *Notophthalmus viridescens*

BODY SHAPE
Sleek; tail flattened side to side

NUMBER OF COSTAL GROOVES
None

BODY COLOR
Brown to olive green

PATTERN
Red spots or broken reddish lines on the back

BODY SIZE
Metamorphs 2 inches; adults up to 4.8 inches in total length

Description Adult eastern newts are brown, greenish brown, yellowish brown, olive green, or nearly black; distinct red spots on the back are surrounded by black or reddish lines bordered by black. The belly is yellow with small black spots. A black line runs horizontally through each eye. During the breeding season the male's tail fins increase in height and black calluses appear on the inside of each hind leg. Females and nonbreeding males have small, almost inconspicuous tail fins and no calluses. Adult newts living aquatically in ponds and other wetlands have smooth, mucous

A black line runs horizontally through each eye.

top Adult eastern newts have distinct red spots on the back bordered by black and a yellow belly with small black spots.

skin; adults that spend a portion of the year in the forest can develop dry, granular skin. The four subspecies differ in color and pattern. Red-spotted newts have red spots on the back, broken-striped newts have partial red lines on the back, and central and peninsula newts lack spots and lines. Peninsula newts are dark olive to nearly black on the back with heavy spotting on the belly.

Larvae and Juveniles Larval newts are greenish with a dark stripe that extends from the snout through the eye to the external gills. The body, head, and tail of the terrestrial juveniles, called red efts, are bright yellowish red, red, or orange-red. Terrestrial adults resemble juvenile red efts in shape and skin texture, but their skin color is light to dark brown rather than red.

top to bottom Red-spotted newt; broken-striped newt; peninsula newt; central newt

Similar Species Striped newts are similar in appearance but have two thin, reddish stripes along the back that extend onto the tail.

left The body, head, and tail of the terrestrial juveniles, called red efts, are brightly colored.

below The dark stripe extends from the snout through the eye to the external gills of a larval eastern newt.

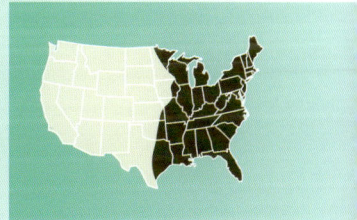

Distribution Eastern newts occur in all eastern U.S. states and from southern Canada to eastern Texas and Oklahoma.

Habitat The aquatic adults and larvae live in freshwater temporary and permanent ponds, lakes, ditches, reservoirs, swamps, bogs, canals, and slow-moving streams, especially where there is abundant vegetation. Fish may be present in the habitat of eastern newts because newts are toxic to them. Efts and terrestrial adults live primarily in hardwood forests but may be found in nearby fields and pine forests.

Eastern newts occur throughout the eastern United States. An adult and juvenile are shown here.

Behavior and Activity

Adult eastern newts can be fully aquatic and remain in the larval pond for the remainder of their lives; or they may migrate to the forest and remain there during the nonbreeding season. If the pond dries up, the aquatic adults emerge and find moist shelter until the next breeding season or until the pond fills again. Mating takes place from late fall through winter in the eastern United States. Larvae are light-sensitive and are active in the water column mostly at night; adults are active day and night. Efts are more commonly found walking on the forest floor at night, especially during and after rains, but they are also active during daytime.

Reproduction

Adults perform an elaborate courtship ritual in water. A male grasps the female behind the head with his hind limbs with the goal of enticing her to pick up a sperm packet with her cloaca. Eggs are fertilized internally with sperm from the packet. The female then lays each of her 50–300 eggs singly on a small leaf or stem of an aquatic plant, sometimes over a period of several weeks. Each egg is wrapped in a leaf and glued shut with secretions from the female's

The male eastern newt at the top is mating with a female. During the breeding season, the males' tail fins become larger.

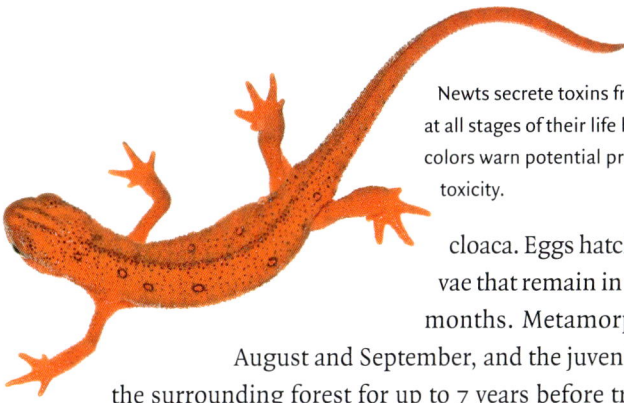

Newts secrete toxins from their skin glands at all stages of their life history. Their bright colors warn potential predators of their toxicity.

cloaca. Eggs hatch into aquatic larvae that remain in the water for 2–5 months. Metamorphosis occurs in August and September, and the juvenile efts remain in the surrounding forest for up to 7 years before transforming into the adult stage and returning to the pond to breed. In some areas of the eastern United States where suitable terrestrial habitat is not available, the larvae skip the eft stage and transform directly into the adult form, possibly retaining their gills as mature adults.

Food and Feeding Larvae eat a wide variety of small aquatic invertebrates, mostly cladocerans, clams, snails, and worms. Terrestrial efts consume adults and larvae of small insects as well as worms, leeches, snails, spiders, and springtails. Aquatic adults eat a wide array of worms, leeches, crustaceans such as grass shrimp, small amphibians and their eggs and larvae, and small fish.

DID YOU KNOW?

No eastern salamanders are venomous (injecting poison into the blood stream) but many produce toxic chemicals on their skin that makes them dangerous if ingested or touched.

Predators and Defense Newts secrete toxins from their skin glands at all stages of their life history that make them unpalatable to most predators. Their bright colors warn potential predators of their toxicity. Birds will not eat the

When picked up or otherwise threatened, newts will often display a peculiar behavior known as the "unken reflex," in which the newt bends its head backward while simultaneously lifting its tail.

brightly colored efts, and fish and most predaceous insects will not eat the adults. Bullfrogs, snapping turtles, painted turtles, garter snakes, and hog-nosed snakes eat an occasional adult but seldom eat the efts, presumably because they are 10 times more toxic than adults. When picked up or otherwise threatened, some newts display a peculiar behavior known as the "unken reflex," in which the newt bends its head backward while simultaneously lifting its tail, forming a circle with its body that prominently displays the brightly colored belly.

Conservation Eastern newts are not federally protected but are declining in many areas as wetlands and surrounding forest habitats are degraded or lost.

Taxonomy Four subspecies are recognized: the red-spotted newt (N. *viridescens viridescens*, described in 1820), the broken-striped newt (N. *v. dorsalis*, described in 1828), the central newt (N. *v. louisianensis*, described in 1914), and the peninsula newt (N. *v. piaropicola*, described in 1952).

STRIPED NEWT   *Notophthalmus perstriatus*

BODY SHAPE	NUMBER OF COSTAL GROOVES	PATTERN	BODY SIZE
Slender body; fleshy tail	None	Irregular series of reddish stripes or small spots on the back and sides	Metamorphs 2 inches; adult up to 4 inches in total length
	BODY COLOR		
	Tan to dark brown		

Description Adult striped newts have rough, tan to dark brown skin. Other characters that may be present include reddish dorsolateral lines thinly bordered by black running the length of the body and extending onto the tail, a row of red dots or streaks on each side of the body, and a faint reddish line along the middle of the back. The belly is yellow with black specks. The back of the cloacal vent of mature males is orange year-round. Males in the breeding season have fleshy tail fins, and their inner thighs have black ridges and pads.

Larvae and Juveniles The aquatic larvae are olive to yellowish green with a black stripe that extends from the snout through the eye to the gills and two irregular, broad dark stripes along each side of the body and tail. Efts are dull tannish brown with reddish stripes on the back and rough skin.

top Striped newts usually have red dots on each side of the body and a yellow belly with black specks.

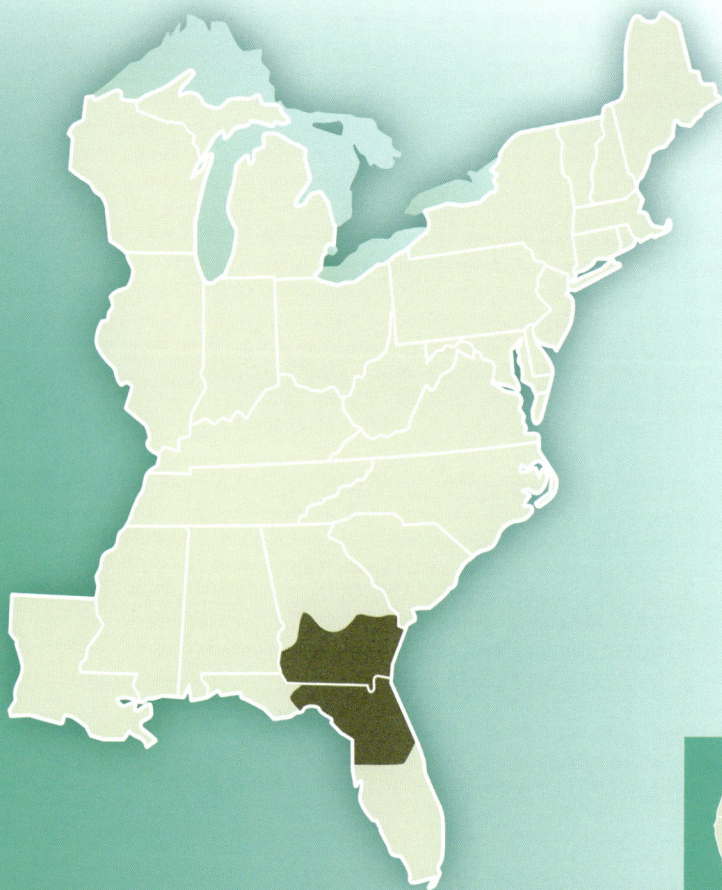

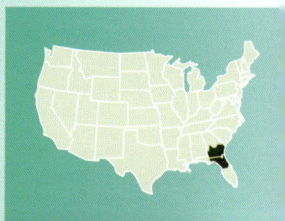

above The aquatic larvae of the striped newt are olive to yellowish green with a black stripe that extends from the snout through the eye to the gills.

left Efts of the striped newt are dull in color with reddish stripes on the back and rough skin.

Similar Species The broken-striped form of the eastern newt has intermittent red striping on the back, and the stripes are bordered boldly by black. Efts of that species usually have dark spots on a yellow to red body.

Distribution Striped newts occur from the Savannah River in southeastern Georgia to north-central Florida, and as far west as the Apalachicola National Forest in Florida and Baker County in southwestern Georgia.

Habitat This salamander occurs in freshwater ponds, limestone sinkhole ponds, and ephemeral natural ponds that lack fish and are on well-drained sandy soils such as those in shrub and sandhill regions. It prefers such aquatic habitats within pine forests and pine-palmetto stands.

Behavior and Activity Adults live in the upland habitats until rains fill the breeding pond, usually in winter and early spring, then migrate hundreds of yards to the pond to mate and lay eggs. Larvae metamorphose in 3–6 months, depending on rainfall and whether wetlands dry up or remain moist. They emerge as efts that spend the next several years in the surrounding forests.

Reproduction The male courts females to encourage them to pick up one of the spermatophores he has deposited on the pond substrate. Sperm from the packet is used to fertilize eggs as they are deposited singly on aquatic plants. The female wraps each egg in a single leaf that she glues shut with her cloacal secretions. The number of eggs laid by a single female is unknown. The larval period usually lasts 3–6 months but may be several months longer. Some populations remain paedomorphic if wetlands do not dry. The larvae mature into gilled adults, skipping the eft stage.

Striped newts occur in southern Georgia and northern Florida.

The striped newt has been proposed for protection under the U.S. Endangered Species Act because the ephemeral wetlands in which it breeds are being lost and upland forests are being converted to pine plantations and agriculture.

Food and Feeding   Larvae eat a wide variety of aquatic insects and their larvae, as well as other invertebrates. Adults eat worms, snails, spiders, frog eggs, and many types of insects and their larvae.

Predators and Defense   No predators have been documented. Toxic secretions from skin glands probably deter most potential predators.

Conservation   This species has been proposed for protection under the U.S. Endangered Species Act because the ephemeral wetlands in which it breeds are being lost and upland forests are being converted to pine plantations and agriculture. Clearing of forest cover around breeding ponds eliminates striped newt populations. Research on restoration of altered habitats and translocation and repatriation into areas where this species once occurred may help enhance the conservation status of this rare species. Currently viable populations and habitats should be protected by modifying silviculture practices. The last Florida surveys conducted on this species indicate that its habitats might be limited to as few as 27 sites.

Taxonomy   The striped newt was first described in 1941. No subspecies are recognized.

forest terrestrial
salamanders

previous page An eastern red-backed salamander

GREEN SALAMANDER *Aneides aeneus*

BODY SHAPE
Flattened head and
body; splayed legs;
large feet

**NUMBER
OF COSTAL
GROOVES**
14 or 15

BODY COLOR
Black

PATTERN
Profuse green
lichen-like pattern
on the head, back,
and tail

BODY SIZE
Hatchlings 0.8
inches; adults up to
5.5 inches in total
length

Description Adults have a flattened head and body, and the long legs have large feet and squared-off toes. The body is black with abundant green to yellowish green markings on the head, back, and tail that resemble lichens. The belly is pale and unmarked.

Larvae and Juveniles Hatchlings and juveniles are colored and patterned as the adults, but the lichen-like, greenish pattern is more diffuse, and more of the black background is visible.

Similar Species The only other salamander with green markings in the range of the green salamander is the Hickory Nut Gorge green salamander (*Aneides*

top This green salamander has the squared-off toes and lichen-like body pattern characteristic of the species.

caryaensis), which has a very limited distribution restricted to the Hickory Nut Gorge of extreme southern North Carolina.

Distribution This species is found in the Appalachian Mountains from Pennsylvania southward through southwestern Virginia and eastern Kentucky to northwestern Georgia and northern Alabama and northeastern Mississippi. Isolated populations occur in western North Carolina, northeastern Georgia, and South Carolina.

Habitat Until recently, the green salamander was thought to be restricted to moist crevices in sandstone, granite, and schist rock formations—habitats the flattened body is well adapted to occupy. Recent discoveries indicate that this

GREEN SALAMANDER *Aneides aeneus*

Green salamanders remain hidden in rock crevices during the day.

species may originally have been an arboreal salamander adapted to living in tall hardwood trees and under bark; large-scale deforestation by humans may have forced it to move to rock crevices. Individuals are occasionally found under rocks and under bark of logs on the forest floor and in moist, human-made structures such as trail bridges well away from rock formations.

Behavior and Activity Green salamanders are nocturnal. They forage in the open on rock faces and trees during rains in all but the coldest winter months. Males defend their rock crevice territories from other males by biting them, pressing down on their back, and flipping them over. Females are not aggressive toward other green salamanders. Both sexes remain hidden in crevices during the day and retreat deeper as surface moisture levels and temperatures decrease.

These nocturnal salamanders forage in the open on rock faces and trees during rains in all but the coldest winter months.

Reproduction Mating occurs in rock crevices and trees primarily in May and June, with a second peak in the fall. Shortly after mating, females attach their 10–27 eggs to the roof of a rock crevice or under bark on fallen logs or even high in trees in small clusters by suspending the eggs with adhesive strands of mucus that they produce while lying on their back.

Green salamanders may originally have been adapted to living in tall hardwood trees and under bark. Large-scale deforestation by humans may have forced it to move to rock crevices

Mothers remain with their eggs until they hatch. All of the development occurs in the egg; there is no free-swimming larva. Hatchlings emerge from eggs after a 3–5-week incubation period and disperse to other crevices or trees with loose bark.

Food and Feeding Adults eat many types of invertebrate prey, including beetles, spiders, flies, crickets, bugs, moths, pseudoscorpions, and mites. Mites appear to be an important prey item for small juveniles.

Predators and Defense Predators include ringneck snakes, garter snakes, and likely some birds that pull these salamanders out of rock crevices. Green salamanders usually remain immobile when disturbed but may wag their tail from side to side at the mouth of the crevice. The potential predator attacks the tail, which the salamander can break off and later regenerate.

Conservation Population declines have been reported for the Blue Ridge Mountain populations, but this species appears to be stable in other parts of its range. Green salamanders may have been more abundant in the first half of the twentieth century before the introduced chestnut blight killed millions of chestnut trees in the early 1900s. While the salamanders may once have lived under the thick bark of large dead chestnut trees, the logs available today are smaller and have thinner bark. The species is listed as Endangered in Mississippi, Protected in Georgia, and Rare in North Carolina, and is a Species of Special Concern in South Carolina.

Taxonomy The green salamander was first described in 1881. Recent research indicates that the green salamander is a species complex comprising as many as four species: northern and southern lineages of *A. aeneus*, the newly described Hickory Nut Gorge green salamander, and a genetically distinct Blue Ridge Escarpment lineage. Ultimately, this complex will probably be designated as four separate species rather than the current two.

HICKORY NUT GORGE GREEN SALAMANDER *Aneides caryaensis*

BODY SHAPE	NUMBER OF COSTAL GROOVES	PATTERN	BODY SIZE
Flattened head and body; splayed legs; squared toes	14	Profuse green, lichen-like pattern on the head, back, and tail	Hatchlings 0.8 inches; adults up to 5.5 inches in total length
	BODY COLOR		
	Black		

Description The basic body color is dark brownish to black with the dorsal surfaces covered by lichen-like patches of bright green to yellowish green. The flanks lack dark pigment, and the belly appears lighter in color than the other surfaces. The head and body are flattened, and the legs are splayed directly lateral of the body. Eyes are large and prominent from head. Females have more teeth than males, which have slightly larger jaw muscles.

Larvae and Juveniles Hatchlings and juveniles have a similar pattern and coloration as the adults. However, their green pattern starts out more diffuse and less organized with more of the dark background visible.

top The head and body of the Hickory Nut Gorge green salamander are flattened, and the eyes are large and prominent.

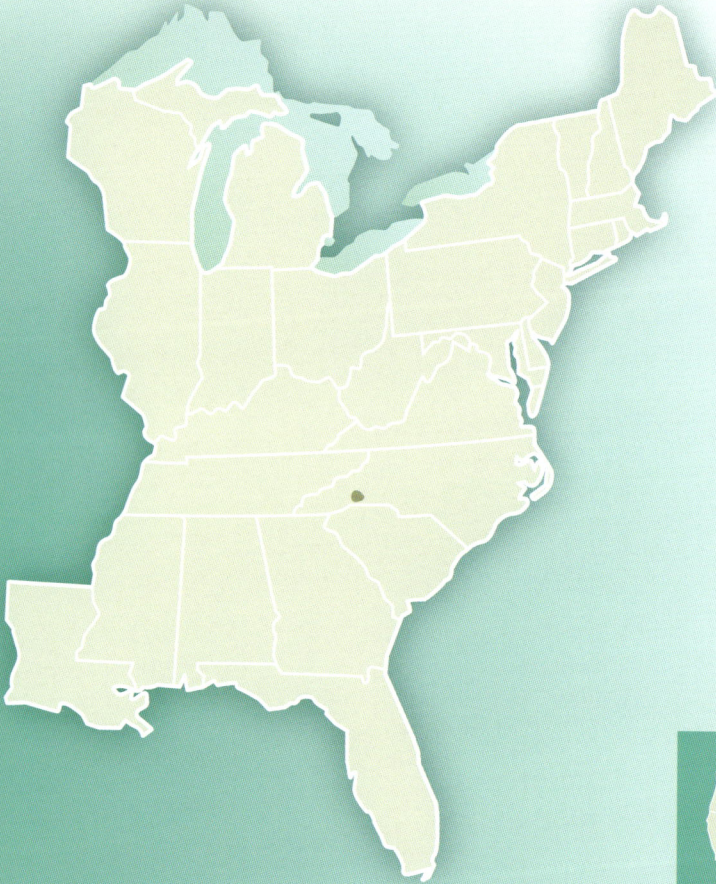

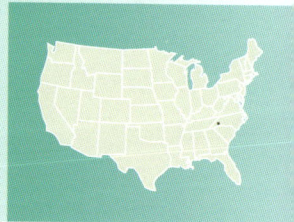

Similar Species No other eastern salamanders have a green pattern except for other species within the *Aneides aeneus* complex. This is the only species in the genus *Aneides* within its limited geographic range. The Hickory Nut Gorge green salamander is slightly smaller than the green salamander and has a longer snout and longer legs and digits. Its background color is more visible with slightly less green coloration.

Distribution The Hickory Nut Gorge green salamander has a limited distribution within the Hickory Nut Gorge area of North Carolina and is known

The Hickory Nut Gorge green salamander has a very limited distribution in the Hickory Nut Gorge area of North Carolina.

from fewer than 25 localities. Preliminary research suggests they are at very low numbers at any one site.

Habitat The Hickory Nut Gorge green salamander is a rock crevice specialist preferring gneiss outcrops in shaded, moist (but not wet) rocks. They use crevices as retreats during the dry and cold periods of the year but do ascend adjacent trees and hide under bark during spring and summer.

Behavior and Activity Hickory Nut Gorge green salamanders are nocturnal in activity and forage on rock surfaces and in trees during all but the coldest months. They are arboreal during much of the year. Males actively defend their rock crevice territories from other males. Females do not appear to be aggressive toward other females. During the day both sexes remain hidden in crevices and retreat into deeper crevices as moisture decreases and temperatures become unsuitable.

Reproduction Like the green salamander, mating probably occurs in May and June with a second peak in the fall. After mating, females attach 10–27 eggs to the roof of a rock crevice or under loose bark on fallen logs or sometimes high in trees. The small cluster of eggs are suspended with strands of adhesive mucus that they produce while lying on their back. Females guard their eggs until they hatch, protecting the eggs from desiccation and small predators. The Hickory Nut Gorge green salamander has direct development with no free-swimming larvae. Hatchlings emerge from eggs after 3–5 weeks of incubation and disperse to nearby crevices or under loose bark on trees.

Food and Feeding Little diet data exists for this newly described species, but its diet is assumed to be similar to that of green salamanders, consisting of a variety of invertebrate prey, including small insects, spiders, pseudoscorpions,

The rare Hickory Nut Gorge green salamanders avoid predation by wedging into tight crevices (*top*) or, if in the open, by relying on their cryptic coloration and remaining still (*bottom*).

and mites. Mites are apparently an important food item for juveniles.

Predators and Defense Predators include ringneck and garter snakes along with spiders and birds. The primary defenses of Hickory Nut Gorge green salamanders are to rely on their cryptic coloration and remain still or to wedge into a tight crevice. When disturbed they will often wag their tails, and sometimes a potential predator will attack the tail rather than its main body. This salamander can break off its tail and later regenerate it.

Conservation Hickory Nut Gorge green salamanders face many conservation concerns and are in need of both protection and habitat management. First, they have a very limited geographic range with scattered suitable rock outcrops. Preliminary studies indicate low numbers and a high amount of inbreeding within the small population. Habitat loss and habitat fragmentation resulting from proximity to tourist and real-estate locations in Hickory Nut Gorge could have a potentially severe impact on the long-term survival of this species. Without doubt this isolated species will require both state and likely federal protection to ensure the protection of its critical habitat.

Taxonomy The Hickory Nut Gorge green salamander was described in 2019 based on both morphological and molecular studies and analyses. It was formerly considered to be a green salamander (*A. aeneus*), but research indicates that the green salamander is a species complex comprising as many as four species: northern and southern lineages of *A. aeneus*, the newly described Hickory Nut Gorge green salamander, and a genetically distinct Blue Ridge Escarpment lineage. Ultimately, this complex will probably be designated as four separate species rather than the current two.

RED HILLS SALAMANDER *Phaeognathus hubrichti*

BODY SHAPE	NUMBER OF COSTAL GROOVES	PATTERN	BODY SIZE
Elongated body; short legs	20–22	Solid body color but paler around mouth	Adults up to 10 inches in total length
	BODY COLOR Dark brown		

Description These long-bodied salamanders have short legs and a thick, prehensile tail; the total length of adults averages more than 6 inches and can surpass 10 inches. A paler brown area around the snout and head and a paler color on the bottom of the feet are the only contrasts with the uniformly dark brown body. The light line from the eye to the angle of the jaw characteristic of the closely related dusky salamanders is absent.

Larvae and Juveniles The eggs develop directly, with no aquatic larval stage, into terrestrial salamanders that resemble adults in body shape and color.

Similar Species All of the dark-bodied salamanders found in the geographic range of the Red Hills salamander have 16 or fewer costal grooves; the Red Hills

top Red Hills salamanders, found only in Alabama, can reach more than 10 inches in total length. They have a dark brown body with no pattern and a thick tail.

The Red Hills salamander spends most of its time underground and is most often observed when sticking its head out of a burrow.

salamander has at least 20. The light line behind the eye that is present in all *Desmognathus* species is absent in the Red Hills salamander.

Distribution

As its common name suggests, this species occurs in the Red Hills physiographic region of southern Alabama. It is found in several counties between the Alabama and Conecuh

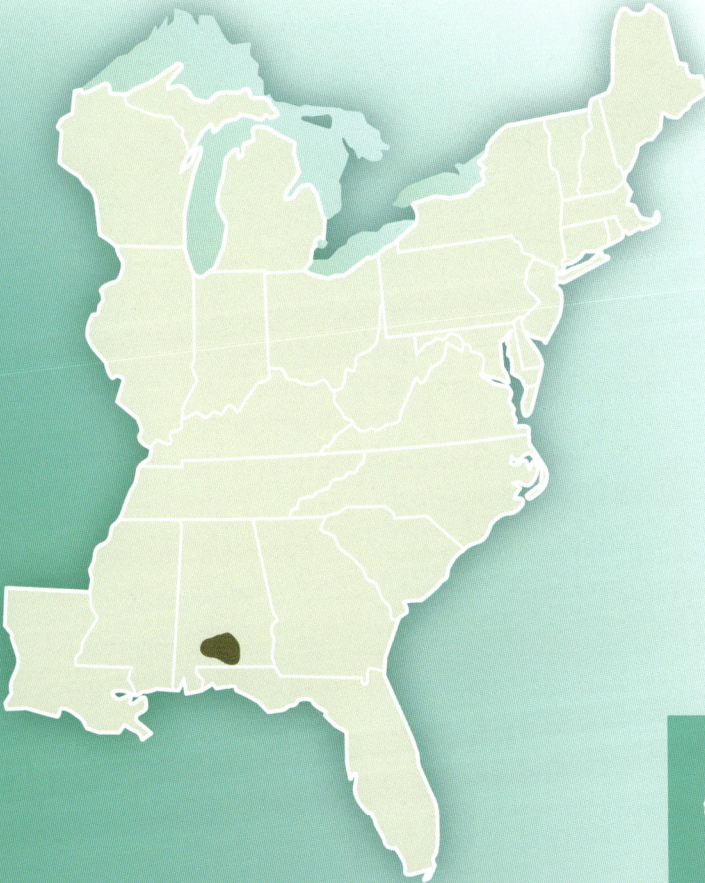

rivers surrounding Butler County, extending in the southeast to Covington County and in the north to Lowndes and Montgomery Counties.

Habitat Hardwood forests with heavy canopies and steep ravines are characteristic habitats, and burrows are more prevalent on the cooler north-facing slopes.

Behavior and Activity The Red Hills salamander spends most of its time underground and is most often observed when sticking its head out of a burrow. More rarely, individuals are seen at night foraging on the ground a short distance from the burrow. Spring and summer are the periods of greatest activity around burrows.

Individual Red Hills salamanders are occasionally seen at night foraging on the ground a short distance from the burrow.

The restricted geographic range and specific habitat requirements of the Red Hills salamander make the species highly sensitive and susceptible to forestry practices that degrade hardwood ravine habitats.

Reproduction Courtship, mating, egg laying, and hatching probably occur underground but have not been observed in nature. The seasonal timing of reproductive events is likewise uncertain. The only information on clutch size is from a captive female that laid 16 eggs. There is no aquatic stage, so all embryonic development must occur within the egg.

Food and Feeding These salamanders feed on insects, spiders, snails, and millipedes, ambushing them from the burrow entrance or finding them on the ground nearby at night.

Predators and Defense No predation has been documented, but small snakes and shrews are likely to eat juveniles and adults. Feral pigs and armadillos that dig up burrows are potential predators.

Conservation The Red Hills salamander is listed as Threatened under the U.S. Endangered Species Act. It is the official state amphibian of Alabama and is protected by that state as a Species of Highest Conservation Concern. Its restricted geographic range and specific habitat requirements make the species highly sensitive and susceptible to forestry practices such as clear-cutting and to other activities that degrade hardwood ravine habitats.

Taxonomy The Red Hills salamander was described as the only member of a new genus in 1961 on the basis of a single specimen discovered in Butler County, Alabama, in 1960. Additional specimens and populations since discovered in the region confirm its uniqueness. No subspecies have been described.

DIXIE CAVERNS SALAMANDER *Plethodon dixi*

BODY SHAPE
Slender and
medium-sized

**NUMBER
OF COSTAL
GROOVES**
15 well-developed
costal grooves;
1 indistinct groove
in the groin

BODY COLOR
Dark gray to black

PATTERN
Heavily marked with
gold to brassy fleck-
ing; white spots on
back and sides

BODY SIZE
Adults up to 4.2
inches in total
length

Description The Dixie Caverns salamander is smaller and slenderer than the closely related Wehrle's salamander (*Plethodon wehrlei*). The body is gray to black and covered with brassy to gold flecking. White spots are numerous on both the back and sides. The belly is light gray with many white flecks; the underside of the chin and throat are a lighter shade of gray (almost white) and mostly lack the white mottling. The hind feet have more webbing than most *Plethodon* species.

Larvae and Juveniles Eggs develop directly into hatchlings that look like miniature adults. Juveniles have more brassy mottling on the back, giving them a more complex dorsal pattern compared to adults.

top Dixie Caverns salamander is an endemic salamander to Virginia with its range centered around Dixie Caverns.

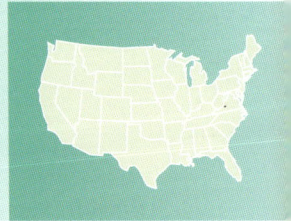

Similar Species Both the Wehrle's salamander and the Blacksburg salamanders look similar to the Dixie Caverns salamander. The Dixie Caverns salamander is mostly restricted to the Dixie Caverns cave system and is smaller and lacks the red spots of the Blacksburg salamander. While Wehrle's salamander does commonly enter caves, it does not seem to overlap the range of the Dixie Caverns salamander and is larger and lacks brassy flecking.

opposite This Dixie Caverns salamander is clinging to the limestone walls within Dixie Caverns.

Distribution The Dixie Caverns sala-mander has a very limited range in central Virginia. It is mostly restricted to the Dixie Caverns cave system (Dixie Caverns and New Dixie Caverns) of Roanoke County. In addition there are several non-cave localities in the vicinity, including Mason Creek and a ridge line near the Roanoke and Montgomery County line.

Habitat Most specimens have been found within Dixie Caverns on the limestone walls where it is damp but not wet. The Dixie Caverns salamander is primarily a cave salamander but is not an obligate troglobite as several small populations have been found in terrestrial forest habitats.

Behavior and Activity Little is known about the Dixie Caverns salamander as it lives in a privately owned cave to which access has been limited. It is speculated to have a greater range but remains subterranean most of the year.

Reproduction Little is known about courtship in the Dixie Caverns salamander. Mating probably occurs in the fall and spring with females laying 7–24 eggs in March or April.

Living within caves and underground affords Dixie Caverns salamanders protection from most predators.

Food and Feeding Dixie Caverns salamanders are opportunistic feeders. They eat small invertebrates including crickets, small spiders, isopods, and millipedes.

Predators and Defense Living within caves and underground affords Dixie Caverns salamanders protection from most predators. No specific defense behaviors or predators have been observed.

Conservation The Dixie Caverns salamander was only recently reelevated to species status and does not yet have recognized federal or state protection.

Taxonomy This species was first described in 1949, but a debate as to whether it should be recognized as a species, subspecies, or "morph" placed within *Plethodon wehrlei* ensued for many years. The Dixie Caverns salamander was reconfirmed to be a legitimate species in 2019 based on molecular analysis.

PEAKS OF OTTER SALAMANDER *Plethodon hubrichti*

BODY SHAPE	BODY COLOR	PATTERN	BODY SIZE
Slender; long tail	Dark brown to nearly black	Brassy to greenish flecking	Adults up to 5 inches in total length

NUMBER OF COSTAL GROOVES

19–22

Description Adults are elongated salamanders with a tail approximately the same length as the body. The body and tail are dark brown to nearly black with highly variable amounts of brassy to mosslike greenish flecking. The sides may have small white spots; the belly is plain dark brown to gray.

Larvae and Juveniles Immature individuals, even hatchlings, are colored and patterned as the adults.

Similar Species Red-backed salamanders have a reddish stripe on the back and tail or are nearly black with fine gray speckling on the body and tail; the

top Peaks of Otter salamanders are dark brown to black with brassy flecking and sometimes with small white spots on the sides. The species' distribution is restricted to only two counties in Virginia.

Peaks of Otter salamanders prefer mixed hardwood forests with a canopy cover; deep, moist soils; and herbaceous plant cover.

belly has a salt-and-pepper pattern with white mottling. White-spotted slimy salamanders are black with white spots on the back and sides.

Distribution   Peaks of Otter salamanders are found only in Bedford and Botetourt Counties in the Blue Ridge physiographic province in Virginia. Most of their small range is in the Jefferson National Forest and the Blue Ridge Parkway above 1,500 feet elevation. Some unknown portion of the range extends into adjacent private property.

Habitat   This species prefers mixed hardwood forests with a canopy cover; deep, moist soils; and herbaceous plant cover in spring and summer. A portion of a population will remain in a site from which timber trees were harvested, but the number remaining is highly variable. These salamanders also occur in rhododendron thickets, mixed hardwood and pine forests, and on forested talus slopes.

Behavior and Activity   Peaks of Otter salamanders are territorial, but the tiny home range of individual salamanders is less than about a square yard. Warm rains will bring them to the surface to forage on the forest floor and on vegetation, including the trunks of trees, up to about 3 feet above the ground. They hide under objects such as logs and rocks during the day and go underground when conditions are dry. The best time to see them active is at night. Only a small portion (about 10 percent) of the population is active on the surface at any given time.

Reproduction   Females lay 5–15 eggs in June and remain with their eggs until hatching occurs in August and September. Otherwise, little has been published on this salamander's courtship, mating, and reproductive biology.

Food and Feeding   Peaks of Otter salamanders eat many types of small invertebrates, including termites, springtails, ants, beetles, spiders, and centipedes.

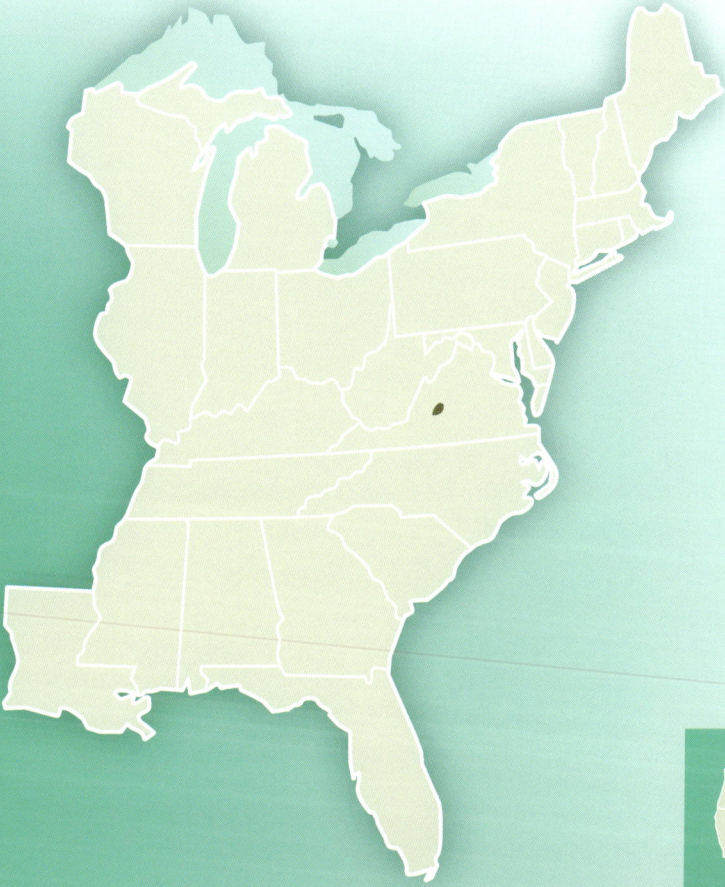

Peaks of Otter salamanders are territorial, but the tiny home range of individuals is only a few square feet.

Only about 10 percent of the population of Peaks of Otter salamanders is active on the surface at any given time, usually at night.

Predators and Defense Ringneck snakes, garter snakes, shrews, and wild turkeys and probably other birds that scrape the leaf litter eat these salamanders. Toxic chemicals in the tail presumably discourage some predators, and remaining immobile when a predator is nearby helps them avoid detection.

Conservation Habitat loss and alteration by removal of hardwood canopy trees appear to threaten populations of Peaks of Otter salamanders, but the population response is variable. The number of salamanders lost depends on how much canopy is removed, how much the forest floor has been disturbed, and whether moist soils persist in the area. A formal federal agency Conservation Agreement and Plan is in place that dictates timber harvest protocols that minimize impacts on this species.

Taxonomy The Peaks of Otter salamander was first described in 1957. No subspecies are recognized.

BLACKSBURG SALAMANDER   *Plethodon jacksoni*

BODY SHAPE
Medium-sized
and slender; large,
webbed rear feet

**NUMBER
OF COSTAL
GROOVES**
15–18; usually 16
or 17

BODY COLOR
Slate gray to blue-
gray

PATTERN
Moderate to heavy
white blotches on
sides

BODY SIZE
Adults up to 5.3
inches in total
length

Description The Blacksburg salamander is a medium-sized salamander with a slate gray to blue-gray body with heavy white blotches on the lower third of its sides sometimes extending onto the belly. These salamanders have moderate to profuse silvery mottling on the back and tail, but the pattern is usually faint or lacking on the head. Red spotting is present on the top of the adults. The belly is variable but normally has a pattern of light and darker areas. The throat is usually lighter than the belly, and often the mid-ventral area is dark gray without white spots. The underside of the tail is slaty-gray.

Larvae and Juveniles This woodland salamander has direct development from the egg and thus lacks a free-living larval stage. Almost nothing is known

top A Blacksburg salamander showing the red spotting typical of adults of this species.

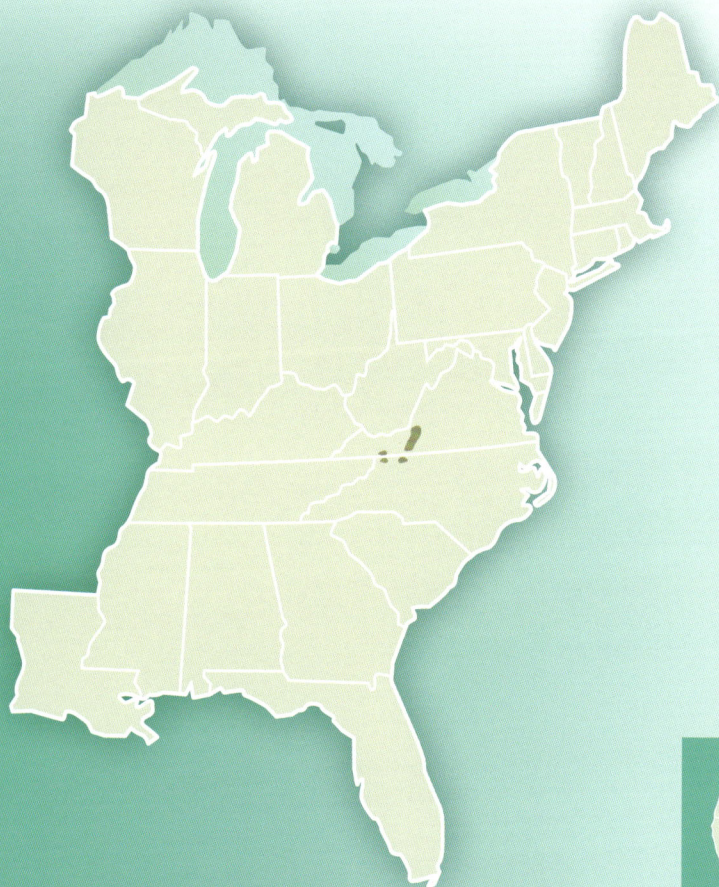

The range of this mysterious salamander is still being analyzed to determine if it is a Blacksburg salamander or possibly a new species. Until recently, salamanders like this specimen from extreme northern North Carolina were classified as Wehrle's salamanders.

The Blacksburg salamander apparently spends most of its life underground within the limestone of the area around Blacksburg, Virginia.

of its eggs or its juvenile stage as apparently these stages of its life cycle occur exclusively underground.

Similar Species The Blacksburg salamander is very similar to its sister species Wehrle's salamander and the Dixie Caverns salamander. The Dixie Caverns salamander has a very restricted distribution, is smaller than either Blacksburg or Wehrle's salamanders, has profuse brassy to golden flecking, and lacks red body spots. The Blacksburg salamander is intermediate in size, has a restricted range, and normally has red spots on its dorsum with 17 costal grooves. The venter of Wehrle's salamanders is darker with less mottling than that of either Dixie Caverns salamanders or Blacksburg salamanders.

Distribution The Blacksburg salamander has a rather small range mainly restricted to Roanoke County, Virginia, although recent research indicates this species may extend south into North Carolina. It is possible, though, that the North Carolina populations constitute a new species.

Habitat Blacksburg salamanders are most commonly found beneath rocks and logs that are partially embedded into the soil. They seem to prefer the drier, upper third of the hillsides. Specimens from Old Mill Cave were found well beyond the twilight zone and quickly react to light. Within caves they can be seen from February to October, but outside of caves they are only found at the surface in late fall, occasionally during warm spells in January and February, then in March through May, completely disappearing by June. They possibly migrate

A Blacksburg salamander from the location where the species was first identified.

vertically through small crevices into the subterranean limestone cavities and then stay there for most of the year.

Behavior and Activity Very little is known about this secretive, elusive salamander. They probably spend much of their life underground, perhaps very deep in fissures and crevices. They are only active in late fall though spring at the surface under cover such as rocks and soil-embedded or decaying logs.

Reproduction Presently nothing is known about the mating and reproduction of Blacksburg salamanders, which probably occurs deep within subterranean cavities.

Food and Feeding Almost nothing is known about the diet of Blacksburg salamanders. Presumably they eat ants, crickets, bugs, flies, beetles, insect larvae, mites, spiders, springtails, earthworms, snails, isopods, and centipedes. On moist or rainy night they forage on the surface and may climb shrubs and trees to search of food.

Predators and Defense Nothing is known about their defense, but ringneck snakes are common within their habitats and are probably predators of the Blacksburg salamander.

Conservation The Blacksburg salamander has only recently been elevated back to the species level of classification and so is not yet included in any federal or state listing of protection. Given its secretive, restrictive range, some level of protection would be recommended to ensure its continued survival and to aid with gaining research funding to better understand and delineate its range and population numbers.

Taxonomy The Blacksburg salamander was originally described in 1954 but was subsequently listed as a subspecies of *Plethodon wehrlei* and then a morph of *P. wehrlei*. Based on evidence from biochemical, ecological, and molecular research, it was reinstated to the species level in 2019.

CHEAT MOUNTAIN
SALAMANDER *Plethodon nettingi*

BODY SHAPE	NUMBER OF COSTAL GROOVES	PATTERN	BODY SIZE
Small and thin	Usually 18	Brassy flecking with small white spots	Adults up to 4.3 inches in total length
	BODY COLOR Black to brown		

Description Cheat Mountain salamanders are small and thin and have brown to black bodies with a moderate amount of brassy flecking, especially on the head. Small white spots may occur on the dorsum. The belly is gray to black.

Larvae and Juveniles The Cheat Mountain salamander has direct development from its egg, with no aquatic larval stage. Hatchlings have some red pigmentation but lack dorsal or lateral stripes. They reach a total length of 0.66–0.70 inches.

Similar Species The Peaks of Otter salamander and Shenandoah salamader look very similar to the Cheat Mountain salamander and were considered subspecies until their separation in 1979. Morphologically they are similar to Cheat

top The rare Cheat Mountain salamander is federally protected and endemic to West Virginia.

Mountain salamanders, but their ranges do not overlap. The Valley and Ridge salamander has 20–21 costal grooves and a mostly white throat. The leadback phase of the eastern red-backed salamander is also similar, but it has a salt-and-pepper pattern on its belly.

Distribution This small salamander has a very restrictive range, confined to the high elevations of Cheat Mountain and nearby mountains in Grant, Pendleton, Pocahontas, Randolph, and Tucker Counties in eastern West Virginia. There are approximately 81 known, geographically isolated populations.

Habitat The Cheat Mountain salamander is often found in red spruce–yellow birch forests with moist ground cover consisting of moss and liverworts (genus

CHEAT MOUNTAIN SALAMANDER *Plethodon nettingi*

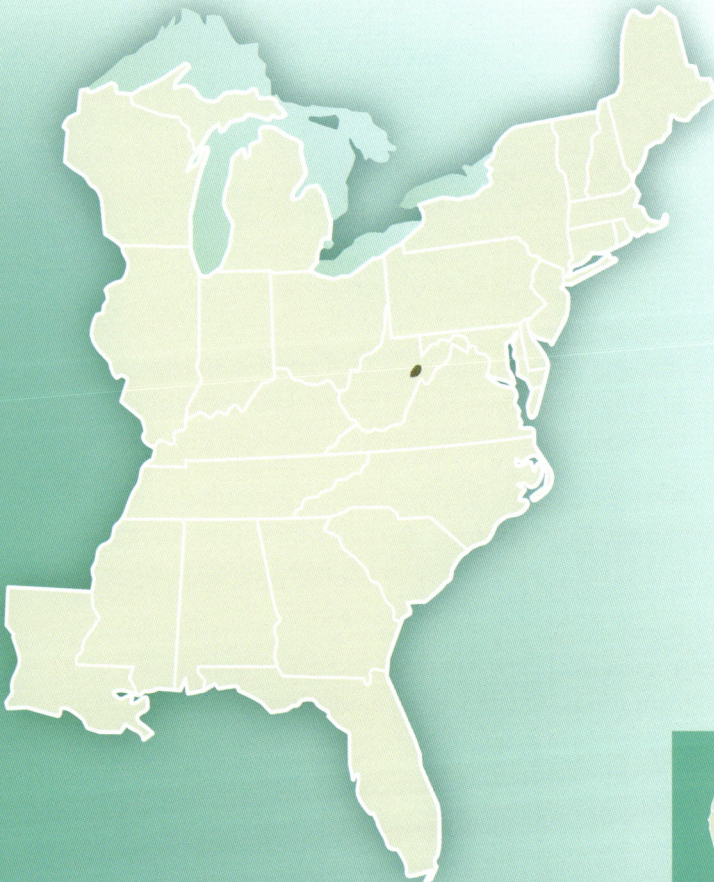

This secretive salamander, before human disturbances, probably lived in high elevation mature red spruce forests.

Bazzania). They have been collected in mixed deciduous forests and are often most abundant in younger spruce stands.

Behavior and Activity These salamanders spend most of their time underground, but surface-active individuals can be found under rocks, within rotting logs, and under logs on humid or rainy nights. They become active after snowmelt in April and remain at the surface throughout the summer unless it becomes dry and hot. Surface activity declines in October with most individuals moving underground by November. Individuals will climb tree trunks and limbs up to nearly six feet above the ground.

Reproduction Egg masses have been found from late April to late August. Females deposit eggs within rotting log cavities or under rocks, with 4–17 eggs suspended by a short pedicel. An adult guards the eggs and very rarely two adults. Freshly laid eggs are pale yellow and approximately 0.16 inches diameter. A female was observed tending eight hatchlings on September 24, 1976.

Food and Feeding The diet of Cheat Mountain salamanders consists of mites, springtails, beetles, flies, ants, and other small invertebrates.

Predators and Defense Natural predators are garter snakes and also probably shrews, ringneck snakes, and ground-foraging birds.

Conservation Due to its very restricted range and isolated, fragmented populations, the Cheat Mountain salamander was federally listed as a Threatened

The greatest threats to the Cheat Mountain salamander are land management practices that remove canopy cover (ski slopes, utility lines, roads, etc.), wildfires, pollution, trails that remove ground cover, and climate change.

Species in 1989, and a recovery plan for the species was adopted. Some isolated populations need further protection, but fortunately 92.6 percent of the species' known range is contained within public lands, primarily those of the U.S. Forest Service. The greatest threats are land management practices that remove canopy cover (ski slopes, utility lines, roads, etc.), wildfires, pollution, trails that remove ground cover, and climate change.

Taxonomy Although first described in 1938 *P. nettingi* was not accepted as a full species until 1979. It had previously been lumped with *Plethodon hubrichti* (Red Hills salamander) and *Plethodon shenandoah* (Shenandoah salamander).

YELLOW-SPOTTED WOODLAND SALAMANDER   *Plethodon pauleyi*

BODY SHAPE	**NUMBER OF COSTAL GROVES**	**BODY COLOR**	**BODY SIZE**
Medium-sized and thin	16–17	Black to lead-colored	Adults up to 6.7 inches in total length
		PATTERN Two rows of yellow dorsal spots	

Description   The yellow-spotted woodland salamander is a medium-sized salamander of the *Plethodon wehrlei* group. The back is black to gray and marked with two rows of yellow, round but irregularly edged spots, each about 0.04–0.08 inches in diameter. The spots (6–12 per side) run from just behind the front legs to just posterior to the hind limbs. The belly is grayish and translucent with lightly pigmented mottling. The chin or gular region is noticeably lighter in coloration than the rest of the belly.

Larvae and Juveniles   This species has direct development like other *Plethodon* species and therefore has no free-swimming larval stage. Juveniles look like miniature adults.

top The yellow-spotted woodland salamander has been recognized for years but only recently described as a distinct species.

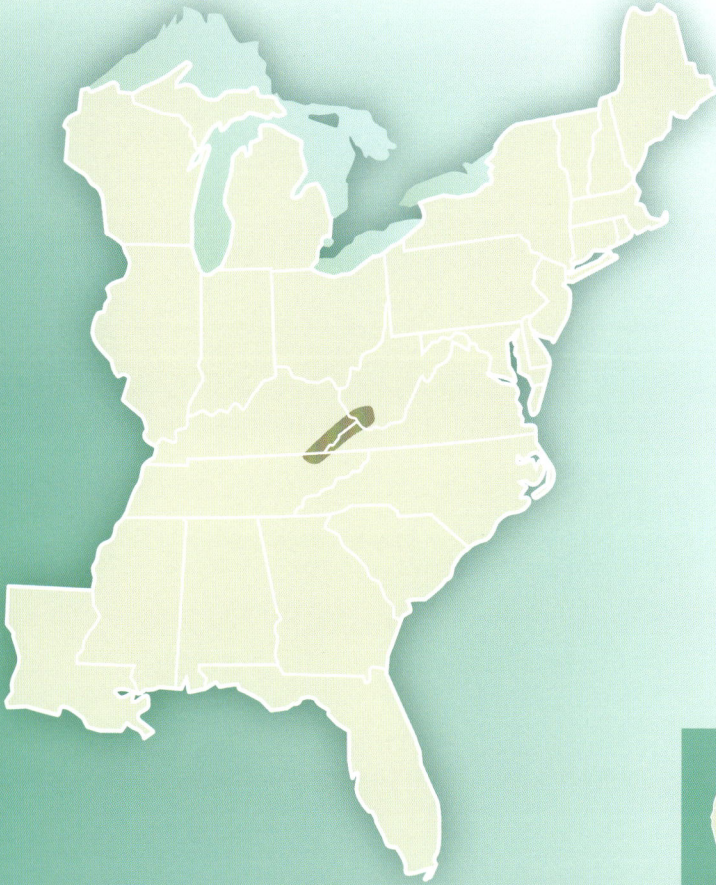

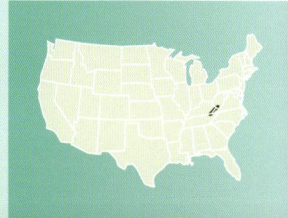

This neonate yellow-spotted woodland salamander is from Tennessee, in the southernmost part of the species' range.

This attractive, crevice-dwelling species is a rock-outcrop specialist.

Similar Species The yellow-spotted woodland salamander could possibly be confused with other species of the *P. wehrlei* group (Dixie Caverns salamanders, Wehrle's salamanders, Blacksburg salamanders, and Cow Knob salamanders) but can be distinguished by its yellow dorsal spots, lightly pigmented belly, and longer limbs.

Distribution This species is only known from a few localities in West Virginia, Kentucky, Virginia, and Tennessee, all within the Central Appalachian ecoregion south of the New River. In the south of their range they are mainly associated with Pine Mountain on the eastern edge of the Cumberland Plateau. Northern populations are found in the New and Bluestone River Gorges. Most of these populations appear small and disjunct from each other.

Habitat The yellow-spotted woodland salamander is apparently a rock-outcrop specialist found on shale or sandstone rock faces. Most have been observed at night climbing in the open or during the day in rock fissures and crevices. A few specimens have been observed at the base of the outcrops on leaf litter or rock surfaces but never far from a rock outcrop.

Behavior and Activity This terrestrial species is most often observed at night but is not common at any location. Little is known about this secretive species' life history.

Reproduction Little is known about reproduction in the yellow-spotted woodland salamander, although it apparently matures at a smaller size than Wehrle's salamander. Females with yolked eggs have been found as small as 2 inches in snout-to-vent length.

Food and Feeding Little diet data exists for the yellow-spotted woodland salamander, but it presumably eats a variety of invertebrate prey such as small insects, spiders, pseudoscorpions, and mites

Predators and Defense Predators are unknown for this species but probably include woodland snakes, shrews, and small ground-foraging birds. Their main defense is seeking refuge within the cracks, fissures, and crevices of rock outcrops.

Conservation Given the limited distribution and specific habitat requirements of the yellow-spotted woodland salamander, it should be considered a species of special concern in need of further research. It occurs in a region where surface mining is common. Kentucky and Tennessee have listed it as Critically Imperiled.

DID YOU KNOW?

The yellow-spotted woodland salamander (Plethodon pauleyi) was named to honor Tom Pauley of Marshall University in West Virginia. Dr. Pauley mentored over ninety graduate students during his career, almost all in the field of herpetology.

Taxonomy P. pauleyi was described in 2019, although it was discovered in 1983, when it was considered a variant of P. wehrlei. The combination of morphological differences, ecological specialty, and genetic divergence supported the elevation of this form to the species level. The P. wehrlei complex has been shown to be highly diverse and has been split into several species (P. punctatus, P. dixi, and P. jacksoni). Current research suggests another split between the northern and southern populations of the currently recognized P. wehrlei is likely.

PYGMY SALAMANDERS

Pygmy Salamander   *Desmognathus wrighti*

Northern Pygmy Salamander   *Desmognathus organi*

BODY SHAPE	NUMBER OF COSTAL GROOVES	PATTERN	BODY SIZE
Slender with a short tail	13 or 14	Wide brownish stripe with a herringbone pattern	Northern pygmy salamander adults 1.57–2.36 inches, with most over 1.97 inches; pygmy salamander adults 1.38–2.16 inches, with few over 1.97 inches.
	BODY COLOR Light brown to yellow-brown to brick red		

Description   Pygmy salamanders are aptly named because adults of these species do not exceed 2.4 inches in total length. The broad, reddish brown to tan stripe along the back is embedded within a herringbone pattern of black chevrons that breaks up on the tail. The sides below the stripe are silvery. These salamanders have a light line that runs from the eye to the angle of the jaw, as do most members of the genus *Desmognathus*. The belly is light tan, and on animals in the southern part of the range (*D. wrighti*), it has a pattern of iridescent

top In addition to their tiny body size, pygmy salamanders, like this example from Rabun Bald, Georgia, have distinctive coppery eyelids.

golden ellipses, whereas northern pygmy salamanders (*D. organi*) either lack or have an incomplete iridescent ventral pattern. The top of the head is rough, and the tail is rounded in cross section (not keeled) and makes up less than half the total length. The copper-colored eyelids are a distinguishing character. Adult males have a U-shaped mental gland. Northern pygmy salamanders are longer, more robust, and have a proportionally longer tail and wider heads than the southern species. Female northern pygmy salamanders tend to be heavier and have slightly longer tails than do the males, whereas the southern species does not show sexual dimorphism.

Larvae and Juveniles Hatchlings have four or five pairs of light spots on the back and a short tail. Juveniles exhibit adult coloration and patterns. There is no aquatic larval stage.

Similar Species Pygmy salamanders often resemble juveniles of other species of dusky salamanders. The top of the seepage salamander's head is smooth, and the mental gland of males is kidney-shaped. Southern red-backed salamanders lack the eye-to-jaw stripe characteristic of dusky salamanders and do not have the copper-colored eyelids of pygmy salamanders. Ocoee salamanders lack golden ellipses on the belly, and the stripe on the back forms a zigzag pattern.

Distribution Due to their preference for higher elevation habitats, pygmy salamanders have one of the most geographically fragmented distributions among members of the genus *Desmognathus*. The northern pygmy salamanders is restricted to high elevations (usually above 3,600 feet) north of the French Broad River valley northward to Whitetop Mountain and Mount Rogers in extreme southwestern Virginia. Its range includes the Black Mountains and Grandfather Mountain in North Carolina and Roan Mountain in Tennessee. The pygmy salamander (*D. wrighti*) is found south of the French Broad River valley and occurs in greatest abundance at high elevations (usually above 4,600 feet) within the Blue Ridge physiographic province. Its range includes the Great Smoky Mountains, the Plott Balsams, and the Great Balsam Mountains. It can also be found at lower elevations in the Cowee, Nantahala, and Unicoi Mountains of North Carolina. The most southerly locality is in extreme northeast Georgia at Rabun Bald.

Habitat Pygmy salamanders can be found in leaf litter and under rocks and decaying logs in and adjacent to seepage areas in high-elevation spruce and fir forests, although they are also commonly found at lower elevations in hardwood forests in the Great Smoky, Cowee, and Nantahala Mountains. They also use

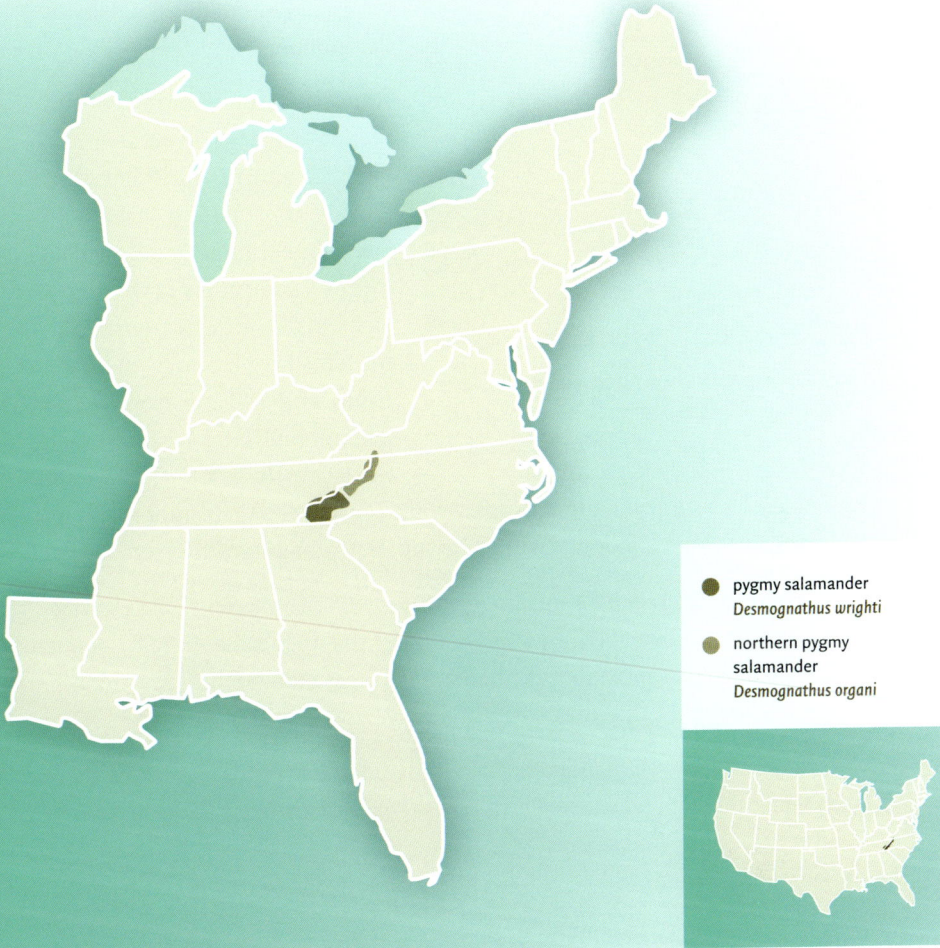

pygmy salamander
Desmognathus wrighti

northern pygmy
salamander
Desmognathus organi

moist forest floor habitats well away from springs and seeps. These salamanders aggregate in winter in seepage areas and stream banks, but in warm weather disperse into the surrounding forest.

Behavior and Activity Pygmy salamanders are active in all but the coldest months. Males disperse from winter hibernacula in late spring and forage in terrestrial sites. Females either remain at their winter sites or disperse to stream banks or seepage areas to nest. Adults and juveniles climb trees and other vegetation, especially ferns, on wet and humid nights to forage for insects and to avoid the larger predatory salamanders on the forest floor. Pygmy salamanders are the most terrestrial of the dusky salamander group.

Reproduction Courtship and mating occur in September and October and in April and May. Shortly after mating, usually in late summer or early fall, female pygmy salamanders lay 8–9 eggs and female northern pygmy salamanders lay 3–10 eggs under the banks of headwater streams and underground retreats. They attach the eggs to the substrate with a single pedicel formed from the egg capsules and remain with them until hatching. All larval development occurs within the egg. Hatchlings have no external gills.

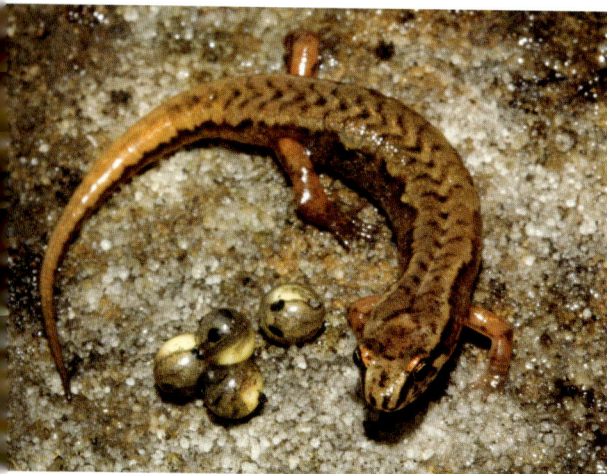

Pygmy salamanders have no aquatic larval stage. This female northern pygmy salamander is guarding her eggs until they hatch into miniature adults.

Food and Feeding Adults and juveniles eat small invertebrates such as springtails, moths, beetles, flies, spiders, and mites that they catch on rainy nights while perched on tree trunks and herbaceous vegetation on the forest floor.

Predators and Defense Spring salamanders and ground beetles are known predators. Pygmy salamanders remain immobile when exposed, making them less noticeable to potential predators. These small salamanders avoid open water and stream banks where larger, predatory salamanders occur.

Conservation The mountaintop spruce and fir forests preferred by pygmy salamanders are greatly reduced as the result of acid precipitation of the 1960s to 1980s. Salamander populations at elevations above 5,000 feet may be declining in part for that reason and possibly also because warmer temperatures and drier conditions resulting from climate change may exceed their tolerance limits. North Carolina, Tennessee, and Virginia recog-

DID YOU KNOW?

Acid precipitation resulting from human industrial activities has killed spruce and fir trees in the Appalachian Mountains, causing population declines in high-elevation species such as pygmy and seepage salamanders.

Northern pygmy salamander adults and juveniles eat a variety of small invertebrates that they catch on rainy nights while perched on vegetation on the forest floor.

nize the pygmy salamander in the category of Special Concern as a result of habitat loss and alteration. The isolated populations in the northern end of the range have less genetic variation and more inbreeding than southern populations, suggesting that they are smaller and more vulnerable to local extirpation and extinction, especially as climate change continues to alter high-elevation habitats.

Taxonomy   The pygmy salamander (D. wrighti) was first described in 1936. Molecular, ecological, and morphological research have revealed two lineages separated by the French Broad River. The northern populations were separated into a new species (D. organi) in 2010, with the southern populations being retained as D. wrighti.

SHENANDOAH SALAMANDER *Plethodon shenandoah*

BODY SHAPE
Slender

NUMBER OF COSTAL GROOVES
Usually 18

BODY COLOR
Black

PATTERN
Uniform or with a reddish stripe

BODY SIZE
Adults up to 4 inches in total length

Description Individuals of this small woodland salamander species either have a narrow reddish to yellowish stripe down the back on a black background or lack the stripe altogether. If present, the stripe extends from the head onto the tail. Individuals that lack the stripe usually have scattered brassy flecking on the back. The belly of both forms is uniformly black, although some

The belly of the Shenandoah salamander is uniformly black.

top Shenandoah salamanders occur in two color patterns. The one shown here has a stripe down the back; the other has no stripe and has a few or no brassy flecks on the back.

individuals have a few scattered light irregular markings. The chin is dark with some scattered flecking.

Larvae and Juveniles Immature Shenandoah salamanders are identical with adults and exhibit the same two patterns—striped or unstriped.

Similar Species Shenandoah salamanders most closely resemble red-backed salamanders, with which they share habitat in some areas. Red-backed salamanders either have a reddish stripe that is broader than that of the Shenandoah salamander or lack the stripe altogether. The belly of red-backed salamanders has a salt-and-pepper pattern, and the chin is uniformly white to cream. Slimy salamanders in the area are black with small white spots on the back and reach a larger size.

Distribution The Shenandoah salamander occurs only on three talus slopes on high mountains (The Pinnacles, Hawksbill, and Stony Man) in the northern part of Shenandoah National Park, Virginia.

Habitat Shenandoah salamanders occupy shallow to deep soil pockets in talus slopes and the margins of the surrounding hardwood forests. Until recently, they also occupied hemlock stands at high elevations (above 3,000 feet), but infestations of the introduced hemlock woolly adelgid bug have killed the hemlocks and eliminated that habitat.

Shenandoah salamanders remain hidden under rocks or logs for most of their lives, emerging only on warm, rainy nights to forage.

Behavior and Activity These salamanders are active in all but the coldest months in winter and dry periods in summer, but they remain under surface rocks or logs most of their lives, emerging only on warm, rainy nights to forage. Adults will defend their rocks from intruders, including red-backed salamanders and other Shenandoah salamanders. Most will not move more than a yard or so from their rock their entire lives.

Reproduction Mating occurs on land in the fall and possibly in early spring. Females lay 4–19 eggs every other year under moist rocks deep in the talus

The range of the Shenandoah salamander is restricted to rocky slopes on mountaintops in Shenandoah National Park, Virginia. The individual shown here shows the uniform, unstriped pattern on its back.

and remain with them until they hatch in 4–6 weeks. Few egg clusters have been found.

Food and Feeding Only their mouth size limits the size of prey these salamanders can eat. The diet includes a wide variety of insects, spiders, centipedes, millipedes, and worms. Arboreal insects that alight on the vegetation and rocks on which these salamanders climb on rainy nights are probably eaten as well.

Predators and Defense Direct predation has not been observed in nature. Ringneck snakes and shrews are the most likely predators. Defensive behaviors include remaining immobile when uncovered, producing noxious secretions in the tail, and breaking off the tail, which distracts the predator while the rest of the salamander escapes.

Conservation Scientists believe that Shenandoah salamanders are restricted to suboptimal habitat in soil pockets in talus by the more aggressive red-backed salamanders that occupy the rocks in the surrounding moist forest. Only three small, isolated populations are known for this rare species. Droughts and climate warming could cause these populations to shrink and thus become even more severely endangered. This species is listed as Endangered under the U.S. Endangered Species Act and by the state of Virginia. Federal scientists are currently monitoring one or more of the populations.

Taxonomy The Shenandoah salamander was first described in 1967. No subspecies are recognized.

YONAHLOSSEE SALAMANDER *Plethodon yonahlossee*

BODY SHAPE	NUMBER OF COSTAL GROOVES	PATTERN	BODY SIZE
Robust body; stout tail	16	Broad, reddish stripe down the back	Adults up to 8.75 inches in total length
	BODY COLOR Black with gray sides		

Description Yonahlossee salamanders are large and handsome members of the woodland salamander group. The broad, red to chestnut stripe down the back and along the base of the tail is characteristic of this species. The stripe is bordered on each side by a narrow, gray to white stripe. The head and limbs are black with light specks; the belly is dark gray with varying amounts of white mottling. Most of the tail is black.

Larvae and Juveniles Hatchlings and juveniles have four to six pairs of red spots instead of the stripe down the back on a gray body. The spots coalesce with age to form the broad reddish stripe.

top Yonahlossee salamanders are larger than most plethodontids and have a wide reddish stripe down the back.

Young Yonahlossee salamanders have red spots instead of the stripe down the back, but the spots coalesce with age to form the broad reddish stripe seen in adults.

Similar Species Other salamanders in the range of the Yonahlossee salamander, such as the red-backed salamander and the mountain dusky salamanders, may have a reddish or yellowish stripe down the back, but the stripe is bordered on each side by dark pigment—never gray or white. The crevice salamander looks similar to the Yonahlossee salamander but has a more southerly distribution centered around Hickory Nut Gorge, North Carolina.

Distribution These high-elevation salamanders are found on high ridges (above 1,400 feet) in southern Floyd and Pulaski Counties in Virginia, and south through the Blue Ridge Mountains in northeastern Tennessee and western North Carolina. Populations in Henderson and Rutherford Counties on the South Carolina border, first described in 1962 as the crevice salamander, have sometimes been included in the Yonahlossee salamander species designation, but the current state of evidence supports their separation into a distinct species.

Habitat Yonahlossee salamanders prefer virgin and old-growth hardwood forests but may also be found in secondary growth forests with a deep leaf litter layer. They also occur in ravines, on road embankments, on talus slopes, near springs, in pastures, and in rock crevices. When they are not underground in burrows, they seek shelter under logs, limbs, bark, and rocks. Juveniles and adults are found in the same habitats.

Behavior and Activity True seasonal activity is unknown because most of the sites occupied by these salamanders are inaccessible during winter. They probably remain underground until spring temperatures allow movement on the surface without the chance of freezing. They emerge from burrows 1–2 hours after sunset to forage on the forest floor, especially during spring, summer, and fall rains, but remain underground during dry periods in summer. Like other forest-breeding terrestrial salamanders, Yonahlossee salamanders stay within a small area of the forest their entire lives.

Reproduction Courtship and mating are terrestrial, with the male enticing the female to pick up a spermatophore with her cloaca. Mating likely occurs after emergence from hibernation in April or May, and may also occur in August and September. The female lays about 19–27 eggs in spring and early summer in an underground cavity and probably remains with them until they hatch in August and September, but this part of the Yonahlossee salamander's natural history has yet to be observed.

Food and Feeding Juveniles and adults eat a wide variety of prey, with only their size limiting what they can take. The diet includes snails, spiders, millipedes, centipedes, beetle adults and larvae, caterpillars, fly larvae and adults, ants, bugs, pseudoscorpions, and worms. Adults occasionally eat other salamanders.

YONAHLOSSEE SALAMANDER *Plethodon yonahlossee*

Both juvenile and adult Yonahlossee salamanders eat a wide variety of invertebrate prey.

Predators and Defense Predators have not been identified but are likely to include snakes and birds such as crows and turkeys that forage in the forest litter. Like other woodland salamanders, Yonahlossee sala-manders release noxious secretions from glands in the tail that deter some predators. Immobility when first exposed makes them less noticeable to visual predators.

Conservation This mountain species is not protected in any state in which it occurs, although many populations are in national parks and on some U.S. Forest Service lands where collecting is prohibited. Logging for timber outside national parks on Forest Service and private lands destroys the canopy and allows sunlight to dry the forest floor, and such habitat alteration is the primary threat.

Taxonomy This species, described in 1917, was named for Yonahlossee Road on Grandfather Mountain in North Carolina, where it was first discovered. The population associated with the Bat Cave and Hickory Nut Gorge area in Henderson and Rutherford Counties, North Carolina, was described as a separate species, the crevice salamander, in 1962.

Yonahlossee salamanders often have burrows in reddish or chestnut brown soils, making their dorsal markings the perfect camouflage.

CREVICE SALAMANDER *Plethodon longricus*

BODY SHAPE	NUMBER OF COSTAL GROOVES	PATTERN	BODY SIZE
Large; long-legged; rounded tail	16	White spots on sides, head, limbs, and tail; varying amounts of chestnut pigment on back	Adults up to 8.5 inches in total length
	BODY COLOR		
	Black		

Description The crevice salamander is a large terrestrial *Plethodon* with a round tail and long, well-developed legs. The ground color is black with numerous white spots along the sides and also on the head, limbs, and tail. Varying amounts of chestnut pigment may appear along the back. The underside is dark gray with scattered white spots, with less white on the chest, tail, and leg insertions. Males are slightly smaller than females and have a large mental gland that is wider than it is long.

Larvae and Juveniles Crevice salamanders have direct development with no free-swimming larvae. Juveniles are 1.12–1.26 inches from snout to vent and have 3–7 small, chestnut brown spots on the dorsum.

top As its name suggests, this salamander spends much of its life tucked into rock crevices and fissures.

Similar Species The crevice salamander looks similar to the Yonahlossee salamander but has a more southerly distribution.

Distribution The crevice salamander is known only from a small area of southern North Carolina near Bat Cave in the Hickory Nut Gorge area of Rutherford and nearby counties. It is found at elevations between 1,430 and 1,712 feet.

Habitat Crevice salamanders occupy deep crevices and fissures and vertical surfaces of gneiss outcrops that offer a cool and moist (but not wet) environment. They prefer areas of mixed pine-hardwood forest often with a dense shrub zone of rhododendron and mountain laurel.

Behavior and Activity Crevice salamanders are nocturnal, spending days hidden within rock fissures and crevices. They have small territories but when not in their rock retreats will venture onto the forest floor, especially during rains.

Reproduction Courtship and mating are terrestrial with the male courting the female and enticing her to pick up his spermatophore. Mating probably occurs after ground temperatures warm up in April or May but may also occur in August and September. Females lay 19–27 eggs in spring and early summer in a crevice, and it is assumed they guard the eggs until hatching.

Food and Feeding The diet of crevice salamanders consists mainly of ants, spiders, millipedes, snails, and weevils, which in one study comprised 63 percent of their food by volume. Other food items include springtails, mites, beetle

The crevice salamander can have varying amounts of brown on its back from almost a stripe to virtually none.

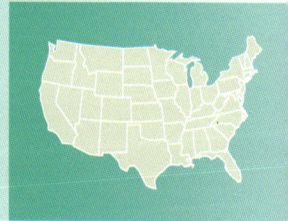

larvae, camel crickets, other crickets, fly larvae, and other insects. Their diet appears less generalized than that of the Yonahlossee salamander.

Predators and Defense Predators have not been identified but likely include small snakes, birds, and shrews. Threatened adults may release noxious secretions from glands in their tail. Their best defense is to stay tucked into a rock crevice and remain still or retreat further in. When disturbed they are quicker to abandon their crevice than the sympatric Hickory Nut Gorge green salamander.

Crevice salamanders have sometimes been grouped as Yonahlossee salamanders.

Conservation The crevice salamander faces many conservation concerns and needs both protection and habitat management. First, it has a very limited endemic range within scattered rock outcrops. Preliminary studies indicate low numbers and a high amount of inbreeding within the small population. Habitat loss and fragmentation resulting from these salamanders' proximity to a tourist and real estate location in Hickory Nut Gorge has impacts on their long-term survival. Without doubt this isolated species requires both state and likely federal protection to ensure that its critical habitat is not lost. Protection and management would also help to protect the other endemic salamander in the area, the Hickory Nut Gorge green salamander.

Taxonomy The crevice salamander was described in 1962, but in 1978 scientists presented evidence that it was the same species as the Yonahlossee salamander. That evidence has been questioned, however, and subsequent research has suggested that the crevice salamander is indeed a separate species. This debate continues, but based on morphological and ecological differences, we chose to present this morph as a distinct species.

GRAY-CHEEKED SALAMANDERS

Blue Ridge Gray-cheeked Salamander *Plethodon amplus*

South Mountain Gray-cheeked Salamander *Plethodon meridianus*

Southern Gray-cheeked Salamander *Plethodon metcalfi*

Northern Gray-cheeked Salamander *Plethodon montanus*

BODY SHAPE	NUMBER OF COSTAL GROOVES	PATTERN	BODY SIZE
Slender; bulging eyes	16	Usually none; rarely white spots and red flecks; gray patch on each side of the head	Adults up to 6 inches in total length
	BODY COLOR Uniform dark gray		

Description The gray-cheeked salamander group includes four species of woodland salamanders formerly included in the red-cheeked salamander (*Plethodon jordani*) complex. All are small (less than 6 inches in total length) with a uniformly dark gray body and gray pigment behind the eye on each side. The Blue Ridge gray-cheeked salamander has a darker belly than the other species in this group. The South Mountain gray-cheeked salamander is larger than the

top This northern gray-cheeked salamander has a solid gray body and tail just like those of the other three species of gray-cheeked salamanders. Genetic analysis and geographic location are the most reliable means of distinguishing among the four species.

others by about 0.5 inches and has a dark belly and light chin. The southern gray-cheeked salamander has a light gray belly in the northern portion of its range and a dark belly in the southern portion; individuals in some southern populations have white spotting on the sides and brassy flecking on the back. The northern gray-cheeked salamander has a light gray belly and even lighter chin. All four species have grayish cheeks and lack red and white pigment; geographic location is the best character to help identify and differentiate among them.

Larvae and Juveniles Juveniles resemble the adults in their respective ranges.

Similar Species Cumberland Plateau, slimy, and southern Appalachian salamanders all have white to brassy spots on a dark body. Yonahlossee salamanders have a broad reddish stripe down the back and tail. The unstriped phases of eastern and southern red-backed salamanders superficially resemble gray-cheeked salamanders, but both forms have a salt-and-pepper-patterned belly.

top left Southern gray-cheeked salamander, *Plethodon metcalfi*

top right Blue Ridge gray-cheeked salamander, *Plethodon amplus*

bottom left Northern gray-cheeked salamander, *Plethodon montanus*

bottom right South Mountain gray-cheeked salamander, *Plethodon meridianus*

- southern gray-cheeked salamander *Plethodon metcalfi*
- both southern gray-cheeked and Blue Ridge gray-cheeked salamanders
- Blue Ridge gray-cheeked salamander *Plethodon amplus*
- northern gray-cheeked salamander *Plethodon montanus*
- South Mountain gray-cheeked salamander *Plethodon meridianus*
- both South Mountain gray-cheeked and southern gray-cheeked salamanders

Distribution Blue Ridge gray-cheeked salamanders are found in the Blue Ridge Mountains in Buncombe, Henderson, and Rutherford Counties, North Carolina. South Mountain gray-cheeked salamanders are found in the South Mountains in Burke, Cleveland, and Rutherford Counties, North Carolina. Southern gray-cheeked salamanders live in the Blue Ridge Mountains in Haywood and Macon Counties, North Carolina; and Oconee County, South Carolina. Northern gray-cheeked salamanders occur in southwestern Virginia from Giles County south through the Valley and Ridge and Blue Ridge physiographic provinces to Haywood County, North Carolina. They occur on Flat Top, Buckhorn, Burkes

The body and tail of southern gray-cheeked salamanders are dark gray. Those in the southern part of the geographic range have a dark belly, and those in more northern areas have a light gray one.

Garden, Knob, Clinch, and Brumley mountain ridges in the Valley and Ridge province; and on Roan, Bald, Black, Max Patch, and Sandymush mountain ridges in the Blue Ridge.

Habitat Gray-cheeked salamanders occupy moist high-elevation, full-canopy forests of mixed oaks, maples, buckeyes, ironwood, gum, and tulip poplar; and forests dominated by spruce and fir. They prefer intact substrates and abundant ground cover in the form of rocks and logs.

Behavior and Activity Gray-cheeked salamanders are active on the surface at night and during rainy weather from mid-May to October but work their way underground as freezing temperatures approach in the fall. Populations at lower elevations may become active on warm, wet days in winter, but high-elevation populations remain underground all winter. Adults and juveniles emerge from their extensive burrow systems on warm, rainy nights to forage. Adults defend territories against intruders of their own species.

Reproduction Courtship and mating take place on land, with each male using visual and tactile behaviors to entice a female to pick up a sperm cap from one of his spermatophores. After fertilization, females lay fewer than 10 eggs, probably in underground cavities although no natural nests have been found. Like other woodland salamanders, females likely remain with their eggs until hatching, which probably occurs in late summer to early fall. The hatchlings remain underground for 10–12 months before emerging on the surface the following May or June.

Food and Feeding Foraging occurs at night on the ground or while the salamander is climbing into herbaceous vegetation and tree trunks as high as 6 feet

This is the so-called Clemson morph of the southern gray-cheeked salamander found in northwestern South Carolina.

above the ground. Known prey includes ants, moth and beetle larvae, moths, beetles, flies and their larvae, wasps, stinkbugs, weevils, crickets, leafhoppers, scorpionflies, gnats, springtails, mites, spiders, millipedes, centipedes, and worms.

Predators and Defense Little is known about predators, but they likely include small snakes, shrews, ground-foraging birds, and other salamanders such as the spring salamander. All four gray-cheeked salamander species have glands in the tail that secrete mucus that may be noxious and may deter some small predators. Immobility following exposure or disturbance may cause a visual predator to overlook the salamander.

Conservation These woodland salamanders are threatened with loss of forest habitat from timber operations, pollution from acid precipitation that kills spruce and fir trees at high elevations, and conversion of forested areas into housing and urban zones. None of these species is listed as Endangered or Threatened.

Taxonomy These four species were considered part of the red-cheeked salamander complex until 2000, when the results of genetic analyses revealed each to be a distinct species.

RED-BACKED SALAMANDERS

Eastern Red-backed Salamander   *Plethodon cinereus*

Big Levels Salamander   *Plethodon sherando*

BODY SHAPE	NUMBER OF COSTAL GROOVES	PATTERN	BODY SIZE
Slender; long tail	18–20, usually 19	Reddish stripe on back and tail, or no stripe	Adults up to 5 inches in total length
	BODY COLOR Dark gray to black		

Description   Red-backed salamanders occur in two color phases in the south-eastern portion of their range: one has a broad, straight-edged, orange, yellow-ish, or reddish stripe on the back and tail (red-backed phase); the other lacks the stripe (lead-backed phase). Both color phases can occur in the same population, although proportions differ geographically. Body color for both is dark gray to black with small white flecks or spots on the head, sides of the body, and tail; and both have a black belly with tiny white or yellowish flecks forming a salt-and-pepper pattern. A uniformly reddish phase of this species with no

top Eastern red-backed salamanders are among the most widespread and abundant salamanders in forested areas of the eastern United States. They are active on the surface and beneath ground litter most of the year.

Big Levels salamanders (*below, left and right*), which are found only in two counties in the Blue Ridge Mountains in Virginia, look like eastern red-backed salamanders (*left*) but have a wider head and longer legs.

stripe occurs in the Northeast and rarely in the Southeast. Big Levels salamanders look like eastern red-backed salamanders but have a wider head, longer limbs, and fewer costal grooves (1–5) between the limbs when they are folded back alongside the body. Eastern red-backed salamanders have 5.5–9 costal grooves between the front and hind limbs.

Larvae and Juveniles Hatchlings and juveniles are identical with the adults of their species but have a shorter tail.

Similar Species Southern ravine and Valley and Ridge salamanders are similar in body color to the lead-backed phase of the eastern red-backed salamander, but their bellies are uniformly black. The orange to reddish stripe on southern red-backed salamanders usually has wavy or saw-tooth edges and extends well onto the tail, but this species is not known to co-occur with the eastern red-backed salamander.

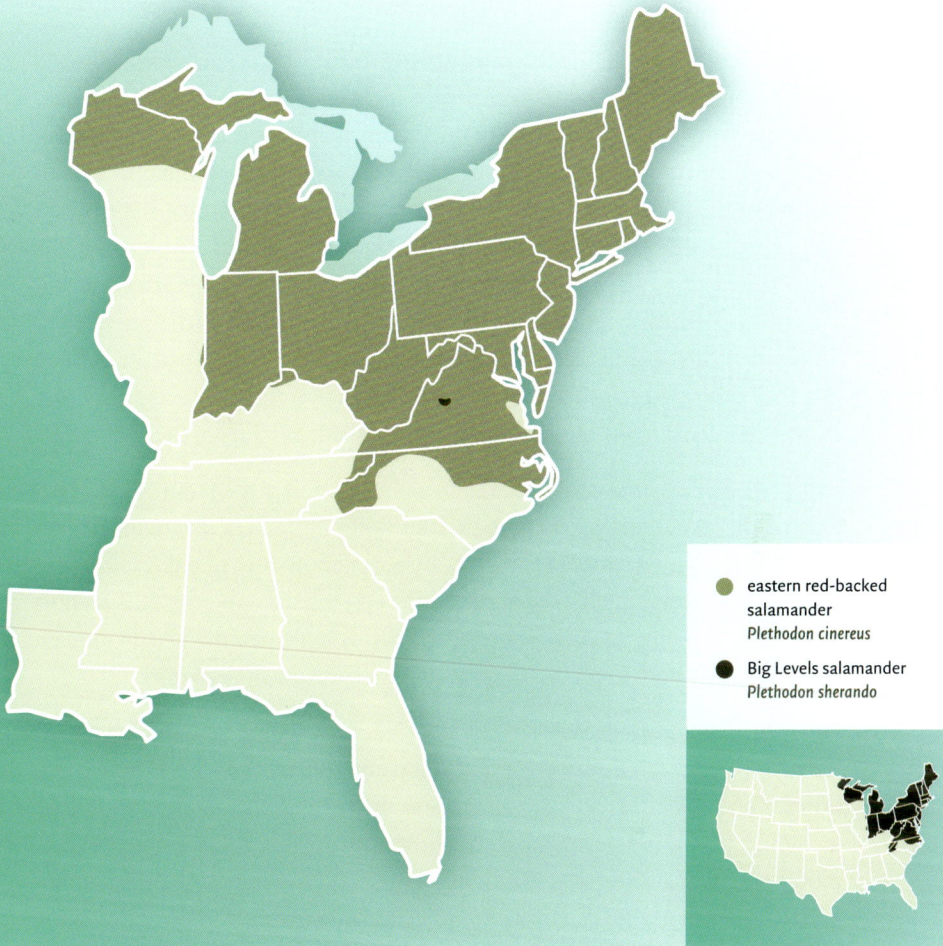

eastern red-backed salamander
Plethodon cinereus

Big Levels salamander
Plethodon sherando

Distribution Eastern red-backed salamanders are widely distributed in north-eastern North America from Minnesota to Maine and south to Kentucky. South-eastern populations occur throughout most of Virginia, much of North Carolina, and the extreme northeastern counties of Tennessee. Big Levels salamanders are limited to a small area in the Blue Ridge Mountains in Augusta and Nelson Counties, Virginia.

Habitat These salamanders prefer deciduous mixed hardwood forests, mixed conifer-hardwood forests, cove forests, and bottomland forests with deep, moist soil–leaf litter layers, where they can reach densities of more than two salamanders per square meter. They occur in lower densities in a wide range of other habitats, including mixed pine and hardwoods, edges of hardwood forests and fields, urban forest patches, and woodlands in agricultural areas.

Behavior and Activity Eastern red-backed salamanders are active in all but the coldest months, although individuals can remain inactive for prolonged periods. Surface activity is greatest during warm rains, when they move about on the forest floor and climb herbaceous vegetation to forage for prey; but only a small portion of the population (2–32 percent) is active on the surface at any given time. They hide under leaf litter, logs, rocks, and other surface objects when the forest floor is moist but work their way down deep into the leaf litter and soil as the surface dries. These salamanders are not active on the surface in hot and dry weather. Males and females defend their small territories, usually defined by a rock or log, from other red-backed salamanders and can recognize other individuals. Most never move more than a few yards away from their birth site.

Reproduction Mating takes place from about November to March. Males court females on land, enticing them to pick up spermatophores. Each female lays 1–14 eggs in a moist cavity in leaf litter, under logs and rocks, inside logs, or in the soil. All development occurs within the egg capsule, and the gills are lost just before hatching in August or September. Female parents usually remain with their eggs until hatching to protect them from small predators.

Food and Feeding These salamanders forage on warm, rainy nights on the forest floor or climb herbaceous vegetation and tree trunks where they sit and wait for prey. The diet consists of any prey they can catch and swallow whole, including ants, mites, termites, earthworms, centipedes, snails, slugs, beetles, moths and their larvae, flies, springtails, and midges. Adults sometimes cannibalize eggs and juveniles. Eastern red-backed salamanders in one population were estimated to eat more than 1.5 million prey items per hectare per year. Their extreme abundance in intact forests ensures a huge role in the ecosystem's energy transfer chain.

Predators and Defense Predators include ringneck snakes, garter snakes, copperheads, shrews, voles, chipmunks, raccoons, foxes, turkeys and other ground-foraging birds, beetles, and praying mantises. Noxious, sticky secretions produced by glands in the tail clog predators' mouths. Red-backed salamanders will also remain motionless to avoid detection, flee, assume a coiled position to protect their head, and break off their tail. The twitching of the broken-off tail distracts the predator while the salamander escapes.

The female red-backed salamander lays as many as 14 eggs in a moist cavity and usually remains with the eggs until they hatch.

Conservation Eastern red-backed salamanders and Big Levels salamanders are not protected in any state in which they occur because they are widespread and abundant. Local populations are threatened, however, when timber harvesting and other practices eliminate the forest canopy and leaf litter, and fragment their habitat into smaller and smaller patches. Because they do not migrate, these salamanders are affected by any activities in their area. Urbanization can cause complete loss of habitat and extirpation of populations.

Taxonomy The eastern red-backed salamander was first described in 1818. The Big Levels salamander was described in 2004 largely on the basis of genetic differences from the eastern red-backed salamander. No subspecies are recognized.

SLIMY SALAMANDERS

Northern Slimy Salamander   *Plethodon glutinosus*

Chattahoochee Slimy Salamander   *Plethodon chattahoochee*

Atlantic Coast Slimy Salamander   *Plethodon chlorobryonis*

White-spotted Slimy Salamander   *Plethodon cylindraceus*

Southeastern Slimy Salamander   *Plethodon grobmani*

Louisiana Slimy Salamander   *Plethodon kisatchie*

Mississippi Slimy Salamander   *Plethodon mississippi*

Ocmulgee Slimy Salamander   *Plethodon ocmulgee*

Savannah Slimy Salamander   *Plethodon savannah*

South Carolina Slimy Salamander   *Plethodon variolatus*

BODY SHAPE
Robust; tail about the same length as the body

NUMBER OF COSTAL GROOVES
Average 16

BODY COLOR
Black to bluish black

PATTERN
White to brassy spots on the back and tail; spots more abundant on the sides

BODY SIZE
Adults up to 8 inches in total length

top Slimy salamanders were once considered a single species but are now divided into 10 different species. The Ocmulgee slimy salamander is pictured here.

top left Chattahoochee slimy salamander
bottom left Northern slimy salamander
right, top to bottom Atlantic Coast slimy salamander;
white-spotted slimy salamander; southeastern slimy salamander; Louisiana slimy salamander; Mississippi slimy salamander; Savannah slimy salamander; South Carolina slimy
salamander

Description

The slimy salamander complex includes ten species found in the eastern United
States that were formerly all considered to be *P.
glutinosus*. All have similar ecology and behavior.
Rely on geographic location or molecular confirmation for species identification.

All slimy salamanders are black to bluish black.
The largest species reach a total length up to 8
inches. Species from the Coastal Plain are at least
half an inch shorter than those from the interior.
Most have white spots or brassy flecks on the back
and tail and a uniformly dark belly. Northern slimy
salamanders have large brass-colored spots on the
back and tail and moderately abundant white and
yellow spotting on the sides of the body. Chattahoochee slimy salamanders have a light chin, little
to no spotting on the back and tail, and abundant
white to yellow spots on the sides. Atlantic Coast
slimy salamanders have small brassy and white

northern slimy salamander
Plethodon glutinosus

both northern slimy and white-spotted slimy salamanders

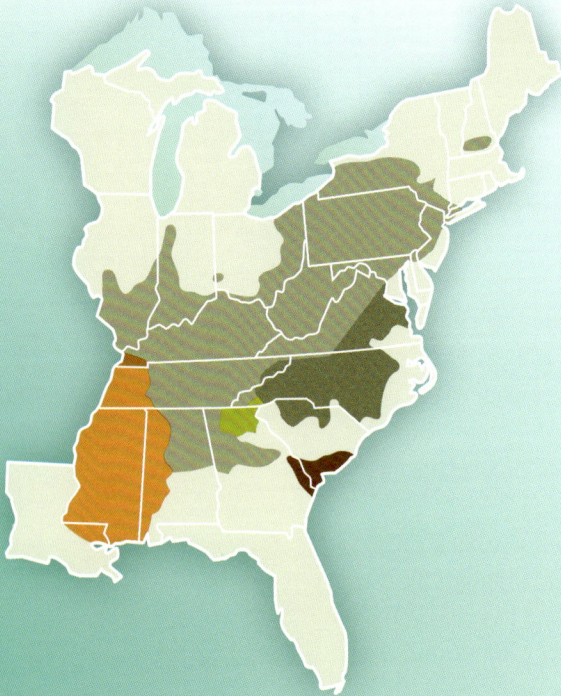

white-spotted slimy salamander
Plethodon cylindraceus

Chattahoochee slimy salamander
Plethodon chattahoochee

South Carolina slimy salamander
Plethodon variolatus

Mississippi slimy salamander
Plethodon mississippi

both Mississippi and northern slimy salamander

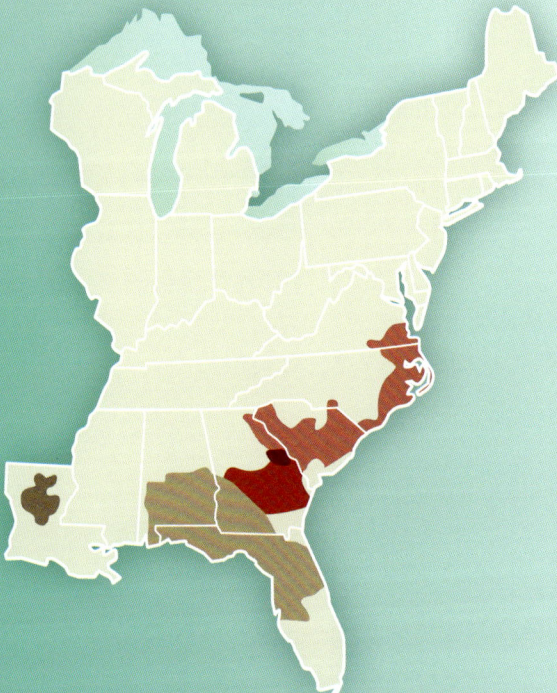

Atlantic Coast slimy salamander
Plethodon chlorobryonis

Savannah slimy salamander
Plethodon savannah

Ocmulgee slimy salamander
Plethodon ocmulgee

southeastern slimy salamander
Plethodon grobmani

Louisiana slimy salamander
Plethodon kisatchie

An adult and neonate northern slimy salamander

spots on the back and tail and abundant white to yellow spotting on the sides. White-spotted slimy salamanders have a light chin, large white spots on the back and tail, and moderately abundant white spots on the sides. Southeastern and Mississippi slimy salamanders have large brassy spots on the back and tail and abundant white to yellow spotting on the sides. Louisiana slimy salamanders have very large brassy to white spots on the back and numerous yellow spots on the sides and legs. Ocmulgee slimy salamanders have few small brassy spots on the back and tail and a moderate amount of white spotting on the sides. Savannah slimy salamanders have very little brassy pigment in the white spots on the back and tail but have abundant white to yellow spotting on the sides. South Carolina slimy salamanders have few brassy spots on the back and tail and a moderate amount of white spotting on the sides.

Larvae and Juveniles Hatchlings and juveniles of each species resemble their parents in color and pattern.

Similar Species Cumberland Plateau salamanders have a light chin, abundant brassy spots on the back, and abundant white to yellow pigment on the sides. Southern Appalachian salamanders resemble slimy salamanders but usually have small red spots on the legs. Tellico salamanders have abundant brassy flecking and spots on the back, but they occur with other slimy salamanders only in Polk County, Tennessee. The lead-backed phase of the eastern red-backed salamander has tiny flecking and a salt-and-pepper belly. Cow Knob salamanders lack brassy or silvery flecking and have small yellow spots on the belly. Some Wehrle's salamanders superficially resemble northern slimy salamanders but have fewer and smaller spots on the back and 17 costal grooves rather than the 16 usually found in slimy salamanders.

Distribution Northern slimy salamanders are widespread from New York to Illinois south through southwestern Virginia, most of Kentucky, all but far western Tennessee, and in northern Alabama and Georgia. Chattahoochee slimy salamanders occur in the northern counties of Georgia that lie in the Blue

Ridge physiographic province and in Cherokee County, North Carolina. Atlantic Coast slimy salamanders are found in the Coastal Plain physiographic province from southeastern Virginia through most of South Carolina into northeastern Georgia. White-spotted slimy salamanders occur in the Piedmont and Blue Ridge physiographic provinces of Virginia and North Carolina to the French Broad River and south into northern South Carolina. Southeastern slimy salamanders occur in southern Alabama and Georgia south into the central Florida peninsula. Louisiana slimy salamanders are found in central Louisiana north into southern Arkansas. Mississippi slimy salamanders are found from western Kentucky south through western Tennessee into Mississippi and the western half of Alabama. Ocmulgee slimy salamanders are known from central and eastern Georgia. Savannah slimy salamanders are known only from eight counties in east-central Georgia. South Carolina slimy salamanders occur in the southern Coastal Plain of South Carolina and in several counties along the Savannah River in eastern Georgia.

Habitat Slimy salamanders occur in a wide variety of habitats throughout their range, including mature hardwood forests, ravines, cove forests, old pine stands, and river floodplains. They use numerous surface objects for shelter on the forest floor, including rocks, logs, slabs of tree trunk or bark, and leaf litter. They are often associated with decomposing tree stumps. Like other woodland salamanders, they remain in a small area of forest their entire lives.

Behavior and Activity Slimy salamanders in the mountains and in Kentucky, Tennessee, and Virginia are active in all warm months of the year—generally from March through October—if rainfall has been sufficient. Those in the Coastal Plain and in the Deep South are more active during winter months than in the hot, dry summer months. They move underground and into burrows and root tunnels as the surface dries, and sometimes form aggregations under large logs. Populations near caves may move inside when the forest dries out and move back out again during rains. They forage

Slimy salamanders, such as this white-spotted slimy, use surface objects, including rocks, logs, slabs of tree trunk or bark, and leaf litter, for shelter on the forest floor.

Adults defend their tiny territories from other slimy salamanders and other woodland salamander species.

on the surface when it is wet or during warm rains. Most move no more than 10 yards from their hatching site. The maximum distance recorded for movement by a juvenile slimy salamander is 92 yards. Adults defend their small territories from other slimy salamanders and other woodland salamander species.

Reproduction Male slimy salamanders court females on land using chemical, tactile, and visual behaviors to entice them to pick up a sperm cap from a spermatophore. Following internal fertilization, females lay 10–33 eggs in clusters that they suspend from the ceiling of a small cavity in the ground, under rocks and logs, and in crevices in caves and

Females lay eggs in small cavities in the ground, under rocks and logs, and in crevices in caves and mine shafts.

mine shafts. Females remain with their eggs until they hatch in August through December following a 2–3 month embryonic development period. The timing depends on latitude, elevation, and proximity to the coast.

Food and Feeding Adults and juveniles emerge from their retreats on wet nights to forage on the forest floor. Appalachian populations are known to feed only in the first few hours after sunset. All slimy salamanders are generalist predators that eat a wide array of prey types. Insect groups include beetles, leafhoppers, aphids, bugs, moths, flies, and ants. Other invertebrate groups include springtails, daddy longlegs, spiders, mites, pseudoscorpions, millipedes, centipedes, worms, snails, slugs, and isopods. They also eat other species of salamanders as well as other individuals of their own species.

Predators and Defense Known predators include copperheads, garter snakes, cave salamanders, and shrews; ground-foraging birds such as turkeys may eat them. The noxious, sticky mucus produced by glands in the tail deters some predators because it tastes bad and glues their mouth shut.

Conservation Hardwood forest habitats used by all slimy salamanders in the group have been subjected to timbering and clearing for agriculture, roads, and

The hardwood forest habitats used by all slimy salamanders have been subjected to timbering and clearing for agriculture, roads, and urban areas.

urban areas. Conversion of hardwood forests to pine monocultures also eliminates habitat. Other threats include alteration of the forest microenvironment by acid precipitation, climate change, and introduced earthworms.

Taxonomy The northern slimy salamander, described in 1818, included all of the other slimy salamander species until molecular analyses determined that the complex comprises a number of forms that should be considered full species. The Chattahoochee slimy salamander, Louisiana slimy salamander, Mississippi slimy salamander, Ocmulgee slimy salamander, and Savannah slimy salamander were recognized as species in 1989. The Atlantic Coast slimy salamander (described in 1951), the white-spotted slimy salamander (described in 1825), and the South Carolina slimy salamander (described in 1818) were resurrected in 1989 from what had been considered incorrect taxonomy.

All slimy salamanders are black to bluish black, and most have white spots or brassy flecks on the back, as seen in the examples pictured here: southeastern slimy salamander (*left*); Mississippi slimy salamander (*bottom left*); and Chattahoochee slimy salamander (*bottom right*).

RIDGE SALAMANDERS

Valley and Ridge Salamander   *Plethodon hoffmani*

Shenandoah Mountain Salamander   *Plethodon virginia*

BODY SHAPE	BODY COLOR	PATTERN	BODY SIZE
Slender; long tail	Brownish black to black	Numerous small white spots or brassy flecks, often concentrated on the sides	Adults up to 5.4 inches in total length
NUMBER OF COSTAL GROOVES			
20 or 21			

Description   The two species can usually be distinguished only genetically or by geographic location. Both are elongated salamanders with short legs and a tail that is longer than the body. Adults are brownish black to black with numerous small, scattered white spots and brassy flecks that are usually more concentrated along the sides. Some red pigment may be present on the back, but most individuals have no hint of a pattern; the exception is some Shenandoah Mountain salamanders on Reddish Knob and Shenandoah Mountain along the Virginia–West Virginia border that have narrow reddish stripes. The throat is white and the belly is black with some white mottling.

top An individual Valley and Ridge salamander can be distinguished from a Shenandoah Mountain salamander only by genetic analysis or geographic location

Larvae and Juveniles Hatchlings have not been described, but juveniles are identical with adults in color and pattern.

Similar Species The lead-backed phase of the eastern red-backed salamander closely resembles both of these species but has a salt-and-pepper belly. Ravine salamanders are even more elongated and have a black belly and light flecking on the throat.

Distribution Valley and Ridge salamanders range from central Pennsylvania south through the Valley and Ridge physiographic province through eastern West Virginia to the New River in Virginia. Shenandoah Mountain salamanders are found in western Rockingham County, Virginia, and northward through

Valley and Ridge salamander
Plethodon hoffmani

Shenandoah Mountain salamander
Plethodon virginia
and Valley and Ridge salamander

Shenandoah Mountain salamanders live in hardwood forests in rocky, mountainous areas, on slopes and in ravines where flat rocks are abundant. They are most active on the surface during moist periods from April through September.

several counties in West Virginia that border Virginia to the Potomac River in Maryland.

Habitat Valley and Ridge and Shenandoah Mountain salamanders occupy deciduous hardwood forests on mountain slopes and ravines where flat rocks are abundant. Individuals can be found under moist logs and rocks and in leaf litter, but most of the population remains underground within the interstices of the rock-soil zone. The substrate in their ranges tends to be dry and well drained. Moist pockets of leaf litter and herbaceous vegetation support greater densities than dry areas.

Behavior and Activity These salamanders are not active on the surface in cold winter months, although they may move about below ground. Surface activity occurs at night during warm rains from April through September, when individuals forage in and under leaf litter and climb herbaceous vegetation to catch prey. They are not active in warm summer months when the ground is dry. Like other small woodland salamanders, they spend most of their lives within a very small area of the forest floor.

Reproduction Mating and courtship take place on land in spring and fall. Males presumably deposit spermatophores and entice females to pick them up, although that behavior has not yet been described for these two species. Females lay three to eight eggs in underground cavities in May and June and likely remain with them until they hatch. Hatching occurs in August and September, but the young do not emerge from underground retreats until the following spring.

The belly of ridge salamanders is black with some white mottling.

Food and Feeding Valley and Ridge and Shenandoah Mountain salamanders consume a wide range of invertebrates including snails, worms, spiders, millipedes, centipedes, collembolans, beetles, flies, ants, and insect larvae. Larger salamanders eat both small and large prey.

Predators and Defense Natural predators have not been observed, but small snakes, shrews, and ground-foraging birds such as turkeys probably eat these salamanders. Both species produce noxious secretions from glands in the tail that deter small predators. They often remain immobile when first uncovered, although some flee quickly to underground retreats.

Conservation These salamanders are not protected in any state in which they occur. Timber management activities in the national forests and private lands in the region alter their habitats by removing canopy trees and exposing the forest floor to sunlight that dries the soil and leaf litter. This may be especially problematic on the already dry soils of the Valley and Ridge physiographic province. Populations are known to decline in areas where the hardwood forests have been removed. Probably the greatest threat to forest floor–dwelling salamanders is the disturbance to the first few inches of soil and humus that occurs during logging.

Taxonomy The Valley and Ridge salamander was first described in 1971, and the Shenandoah Mountain salamander was described as a separate species in 1999 on the basis of molecular differences. No subspecies are recognized.

CUMBERLAND PLATEAU SALAMANDER *Plethodon kentucki*

BODY SHAPE
Slender; bulging eyes

NUMBER OF COSTAL GROOVES
16

BODY COLOR
Black

PATTERN
Scattered white spots that are more concentrated on the sides; light chin and throat

BODY SIZE
Adults up to 6.5 inches in total length

Description Cumberland Plateau salamanders are relatively large and have big, bulging eyes. They are entirely black with small white spots scattered on the head, back, and tail and a greater concentration of larger, irregular white spots on the sides. The belly is uniformly slate gray; the chin and throat are light gray. Males have a very large mental (chin) gland.

top The Cumberland Plateau salamander is superficially similar to the northern slimy salamander, from which it can be differentiated by genetic analysis and geographic location. Its head is also relatively broader and slightly flattened.

bottom Note the light gray chin and throat and prominent mental gland on this male.

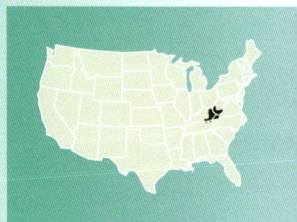

Larvae and Juveniles Hatchlings and juveniles closely resemble the adults but have fewer white spots.

Similar Species The northern slimy salamander is very similar in appearance, and its range completely overlaps that of the Cumberland Plateau salamander. Slimy salamanders are black with an abundance of silvery white to metallic gold spots on the head, body, and tail; larger spots on the sides; and a grayish black belly, throat, and chin. Adult male slimy salamanders have a smaller chin gland.

left Cumberland Plateau salamanders often seek hiding places in rock crevices.

below Adult and juvenile Cumberland Plateau salamanders are active on the surface on warm, moist days.

Distribution Cumberland Plateau salamanders occur in southern West Virginia, the eastern third of Kentucky, southwestern Virginia, and several counties in Tennessee along the Tennessee-Kentucky state line.

Habitat These woodland salamanders prefer mature hardwood forests. Habitats include mixed hardwood forests, moist ravines, hillsides, shale banks, and cave entrances. Individuals use rocks and logs on the forest floor for shelter but often seek hiding places in rock crevices.

Behavior and Activity Adults and juveniles are active on the surface on warm, moist days from about April through October, although they go underground during hot, dry weather. Rain brings them out to forage. They remain in a small area their entire lives; known movement distances are only 1–2 yards.

Reproduction Male Cumberland Plateau salamanders, like other *Plethodon* species, court and mate on land in summer and fall. The male uses visual and tactile stimulation to entice a female to pick up the sperm cap from the spermatophore he has deposited on leaf litter. Once fertilization has occurred internally, females deposit 9–12 eggs in clusters in underground cavities in summer and remain with them until they hatch in September and October. Hatchlings remain with their mother for several weeks until their yolk reserves are used up, then move into underground burrows and crevices on their own. They emerge on the forest floor the following June or July when they are 8–9 months old.

Food and Feeding All size classes forage on the forest floor and on low vegetation, eating anything they can subdue and swallow. Prey items include forest floor dwellers such as ants, springtails, beetles, flies, mites, spiders, pseudoscorpions, and snails.

The Cumberland Plateau salamander was considered to be a slimy salamander until 1983, when it was elevated to full species status due to genetic evidence.

Predators and Defense Predators of Cumberland Plateau salamanders have not been reported, but snakes and ground-foraging birds undoubtedly eat them. Like slimy salamanders, they produce noxious secretions from glands in the skin that deter small predators.

Conservation This species is not protected in any state in which it occurs. Loss and alteration of mature hardwood forests from timbering operations and conversion to agriculture and urbanization cause populations to decline and disappear.

Taxonomy The Cumberland Plateau salamander was first described in 1951 as a full species, but from the mid-1950s until the mid-1980s was thought to be a slimy salamander. Genetic analysis in 1983 showed that the Cumberland Plateau salamander differs from slimy salamanders enough to warrant full species recognition.

SOUTHERN RED-BACKED
SALAMANDER   *Plethodon serratus*

BODY SHAPE
Slender; tail round
and less than half
the total length

**NUMBER
OF COSTAL
GROOVES**
18–22

BODY COLOR
Brown with a small
amount of reddish
and white pigment

PATTERN
Stripe along the
middle of the back
and tail, or no stripe

BODY SIZE
Adults up to 4.2
inches in total
length

Description   Southern red-backed salamanders occur in both striped and unstriped forms. The striped version has a reddish to orangish, or rarely gray,
stripe down the middle of the back and tail. The edges of the stripe are distinctly
serrated, and each small projection alternates with a costal groove. The color
of the stripe often fades onto the sides, especially along the front of the body.
The back and tail of the unstriped form, which is found primarily in Alabama
and Georgia, are brownish to gray; the back, sides, and belly have scattered red
spots. Both forms have brown sides with varying amounts of white. The tail is
round and shorter than the body.

top Southern red-backed salamanders are small and occur in two color patterns; this is the
striped form. All individuals have brown sides, usually with some white markings.

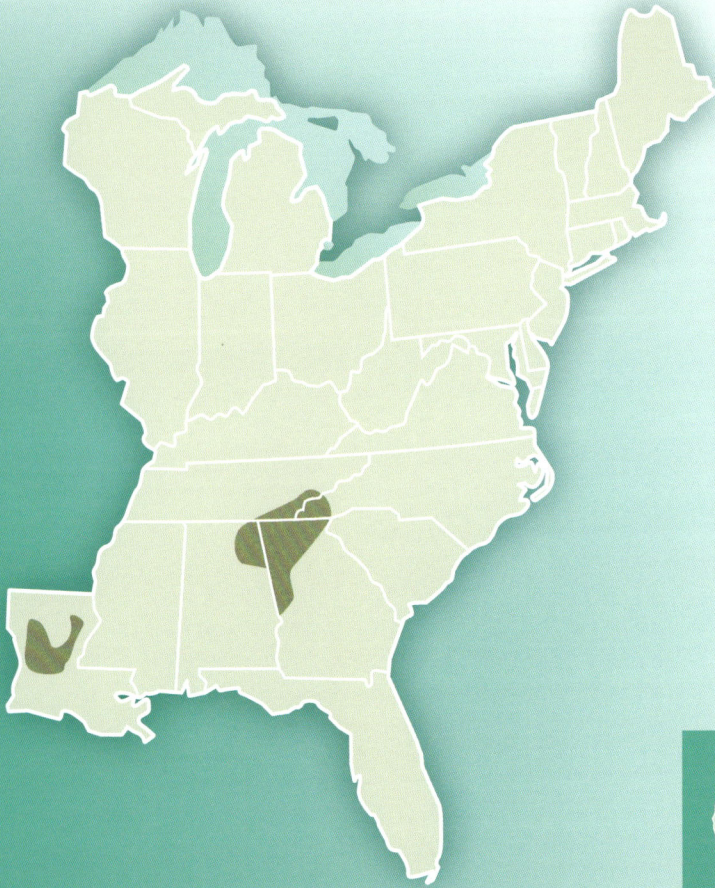

Some southern red-backed salamanders without the reddish stripe on the back (lead-backed phase) also lack the brown color on each side.

Southern red-backed salamanders are found in a wide variety of habitat types, most with one or more species of hardwood trees and abundant ground cover.

Larvae and Juveniles Juveniles are patterned and colored as the adults.

Similar Species Eastern red-backed salamanders are similar in color and pattern but have a salt-and-pepper belly pattern, and the ranges of the two species do not overlap. The reddish stripe on southern zigzag salamanders is distinctly lobed along the edges rather than serrate, especially on the front half of the body. The orange to reddish stripe on Webster's salamanders has wavy edges, at least on the front half of the body. The latter two species also have 18 costal grooves.

Distribution Southern red-backed salamanders occur in four widely separated areas in western Arkansas, Missouri, central Louisiana, and southwestern North Carolina to central Georgia.

Habitat These salamanders occupy a wide variety of habitat types, most with one or more species of hardwood trees and abundant ground cover in the form of logs, rocks, and leaf litter. Habitats include pine-oak forests; mixed deciduous hardwoods such as oaks, yellow poplar, and American beech; cove forests; and bottomland forests. The species has also been found in longleaf pine habitat on rocky hillsides in Louisiana. These salamanders can occupy relatively drier soils than other woodland salamanders but usually remain in moist leaf litter and under objects that retain moisture.

Behavior and Activity Southern red-backed salamanders move only a few linear yards during their entire lives and stay in the same small patch of forest. Movement is mostly from the surface to underground retreats during dry periods. They are seldom found on the surface from May to September but are active during all other months of the year. Juveniles appear in October and adults are first active in November. Like their woodland salamander relatives, these salamanders are territorial.

Adults and juveniles secrete noxious fluids that deter some predators.

Reproduction Courtship and mating are terrestrial and take place in winter. Females lay an average of five eggs deep in the ground in June and July and probably remain with them until they hatch in August and September; hatchlings can be found on the surface as early as October. Too few observations have been published to provide a complete picture of the reproductive cycle.

Food and Feeding Southern red-backed salamanders eat any animal they can subdue and swallow, including snails, worms, mites, spiders, millipedes, centipedes, pillbugs, thrips, flies, moths, crickets, beetles, ants, and the larvae of many insects. Smaller salamanders eat smaller prey; adults eat both small and large prey.

Predators and Defense No predators have been documented, although undoubtedly small snakes, ground-foraging birds, shrews, and some predatory insects eat these salamanders. Adults and juveniles secrete noxious fluids from skin glands that deter some predators. Individuals become immobile when first exposed, decreasing the likelihood that a visual predator will notice them, and may autotomize the tail when attacked.

Conservation Southern red-backed salamander populations decline when hardwood forests are harvested for timber or cleared for roads, fields, and urban development. This species is not listed as Endangered or Threatened by any state in which it occurs.

Taxonomy The southern red-backed salamander was first described in 1944 as a subspecies of the eastern red-backed salamander and was elevated to full species status in 1976 on the basis of molecular evidence. No subspecies are recognized.

WEHRLE'S SALAMANDER *Plethodon wehrlei*

BODY SHAPE
Slender body; large, webbed rear feet

NUMBER OF COSTAL GROOVES
17

BODY COLOR
Dark brown to gray-black

PATTERN
Small white spots, concentrated on sides

BODY SIZE
Adults up to 6.5 inches in total length

Description Wehrle's salamanders are middle-sized members of the woodland salamander group. The body is dark brown or dark gray with small, scattered white spots or brassy flecks on the head, back, and tail. Some individuals have small reddish spots on the shoulders. The sides of the body have abundant bluish white to yellowish spots that often fuse to form blotches or dashes. The belly is uniformly dark gray, but the throat is white or has white blotches. The webbing on the expanded rear feet reaches almost to the tips of the first two toes.

Larvae and Juveniles Hatchlings and juveniles in the northern populations usually lack pairs of red spots on the shoulders.

top Wehrle's salamanders are usually found at higher elevations in moist mixed deciduous and coniferous forests, although they sometimes occur on drier, rocky forested hillsides.

Similar Species Northern slimy salamanders closely resemble Wehrle's sala-manders but have larger and more numerous spots on the back, a gray throat, and 16 costal grooves. Northern gray-cheeked salamanders are uniformly dark gray above and lighter gray below and have gray patches on the sides of the neck behind the head. The closely related Blacksburg and Dixie Caverns salamanders are smaller and have very restricted ranges. The venter of the Wehrle's salaman-der is darker with less mottling than those species, and the dorsum lacks either brassy flecks or red spots.

Distribution Wehrle's salamanders occur from western New York south through Pennsylvania and West Virginia into the mountains of southwestern Virginia. Populations in North Carolina may be P. wehrlei or Plethodon jacksoni (Blacksburg salamanders) or an as-yet-undescribed species of the P. wehrlei complex.

Habitat These salamanders are found at elevations above 600 feet, and primarily on high mountains. They prefer mixed deciduous and coniferous forests such as spruce–yellow birch but also occupy dry, forested hillsides where there are rock ledges and flat rocks. They hide under surface objects such as logs, rocks, and leaves but go underground when the surface is too dry. Individuals are common in some caves.

Behavior and Activity Wehrle's salamanders are active on the surface from March to October at lower elevations, and from May to early October at high elevations. Activity periods are controlled by the timing of rain; salamanders remain underground when temperatures are high and the ground and leaf litter are dry. Like other woodland salamanders, Wehrle's salamanders remain in a small area their entire lives. Adults defend territories—usually including a cover object—on the forest floor from other adults and sometimes other species. They do not migrate but move vertically in response to surface moisture.

Reproduction Mating is terrestrial, with males courting females to entice them to pick up sperm caps from spermatophores deposited on the ground. Mating

DID YOU KNOW?

Species in the Plethodontidae, the largest family of salamanders in the world, have no lungs. All terrestrial salamanders respire through their moist skin.

Wehrle's salamanders are found at elevations above 600 feet, primarily on high mountains. Individuals are common in some caves.

occurs in fall and spring. Females lay 7–24 eggs in underground cavities in hardwood and coniferous forests in March to April. One clutch was found in a cave. Females likely remain with their eggs until they hatch in August to September, as other *Plethodon* females do.

Food and Feeding Adults and juveniles forage for prey on the forest floor and on low vegetation and tree trunks during warm and wet periods. Known prey items include ants, crickets, bugs, beetles, flies, weevils, daddy longlegs, a variety of insect larvae, mites, spiders, springtails, worms, snails, isopods, millipedes, and centipedes.

Predators and Defense The only known predator of Wehrle's salamander is the ringneck snake. Other snakes such as garter snakes, ground-foraging birds, and shrews probably eat them as well.

Conservation Wehrle's salamanders are listed as Threatened in North Carolina; however, the populations in that state will probably be identified as Blacksburg salamanders (*P. jacksoni*) or described as a new species. Clearing of hardwood and spruce forests for timber sales, roads, and houses constitutes the most important threat to populations. Changing temperature and rainfall patterns due to climate change may cause contraction of the geographic range.

Taxonomy Wehrle's salamander was first described in 1917. Dixie Caverns salamanders and Blacksburg salamanders were described as separate species in the twentieth century but were later merged with *P. wehrlei*. Both species were reelevated to full species status in 2019 based on molecular evidence.

WELLER'S SALAMANDER *Plethodon welleri*

BODY SHAPE	NUMBER OF COSTAL GROOVES	BODY COLOR	BODY SIZE
Small with a thick tail	15–17	Black	Adults up to 3 inches in total length
		PATTERN Abundant brassy markings on the body and tail	

Description Weller's salamanders are small and dark gray to black with numerous irregular, gold to brassy blotches on the head, back, and tail that often run together to form large patches. The belly varies from black with numerous fine white spots in most of the range to uniformly black in the vicinity of Grandfather Mountain in western North Carolina.

Larvae and Juveniles Hatchlings have several pairs of brassy spots along the back near the head; the body is otherwise black.

Similar Species The lead-backed phase of the eastern red-backed salamander is similar in color but lacks the heavy brassy pattern. Pygmy salamanders have

top Brassy blotches on the body and tail coupled with a small body size are characteristic of Weller's salamanders, which are found on mountaintops, sometimes at elevations above 5,000 feet.

WELLER'S SALAMANDER *Plethodon welleri*

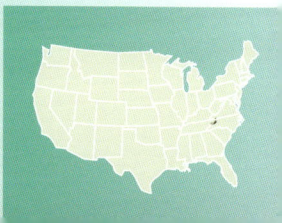

a light stripe down the back and a light eye-to-jaw stripe. Juvenile Yonahlossee salamanders have four to six pairs of reddish spots on the back.

Distribution This small salamander occurs on several isolated mountaintops from Mount Rogers and Whitetop Mountain in southwestern Virginia through the Unaka Mountains in the Blue Ridge in northeastern Tennessee and in northwestern North Carolina south as far as Grandfather Mountain.

Habitat This is a mountaintop species found primarily in spruce-fir forests at elevations above 5,000 feet, and in lower numbers down to about 2,500 feet in moist cove hardwoods. Individuals use logs, rocks, and rock flakes for cover on the forest floor, but spend much of their time under mats of moss and below the surface in leaf and pine needle litter.

Behavior and Activity Weller's salamanders are active from about April or May to October, and are probably inactive near the surface during the cold winter months. These small salamanders defend small territories against intruders, and males attack one another on sight; however, they also aggregate without fighting for long periods during winter.

Reproduction Mating takes place in spring and fall on the forest floor when the male entices a female—through tail waving and touching and rubbing his chin on her back—to pick up the sperm cap sitting atop a spermatophore. Fertilization is internal. Females lay 4–11 eggs in summer in decomposing spruce or fir logs and under moss mats. The mother stays with the eggs until they hatch in early fall.

Food and Feeding Weller's salamanders forage for insects and other invertebrates beneath the forest floor surface and on the forest floor at night during wet periods. Known prey items of juveniles and adults include beetles, moth larvae, bugs, flies, springtails, mites, spiders, and pseudoscorpions. Individuals eat their own skin when they molt.

Predators and Defense Predators have not been reported but undoubtedly include birds that forage on the forest floor and possibly small snakes and mammals found at high elevations. Noxious secretions produced by skin glands in the tail are thought to deter some predators. Individuals also become immobile upon discovery, a behavior likely to make visual predators overlook them.

The loss of the spruce and fir forests in the high elevations of the Blue Ridge to logging and acid rain pollution is a major threat to the long-term persistence of Weller's salamanders.

Conservation The loss of the spruce and fir forests in the high elevations of the Blue Ridge to logging and acid rain pollution is a major threat to the long-term persistence of this salamander because the forest floor environment is deteriorating. The introduced balsam woolly adelgid has caused serious damage to Fraser fir forests. Increasing temperatures predicted by global climate models may further restrict the range of this species to high mountaintops and could eliminate some lower-elevation populations altogether. Weller's salamander is listed as a Species of Special Concern by North Carolina and is protected by Tennessee as Wildlife in Need of Management.

Taxonomy This salamander was described in 1931 and named for Worth Hamilton Weller of Cincinnati, Ohio, who died collecting it on Grandfather Mountain in 1931 on a field trip at the age of 18.

BAY SPRINGS SALAMANDER *Plethodon ainsworthi*

BODY SHAPE
Slender

NUMBER OF COSTAL GROOVES
16

BODY COLOR
Dark brown or black

PATTERN
None

BODY SIZE
Adults up to 4 inches in total length

This species is believed to be extinct. No information is available on behavior and activity, reproduction, food and feeding, or predators and defense.

Description Based only on preserved specimens, the Bay Springs salamander is slender, has reduced limbs, and is dark brown or black with no visible pattern.

Larvae and Juveniles Unknown.

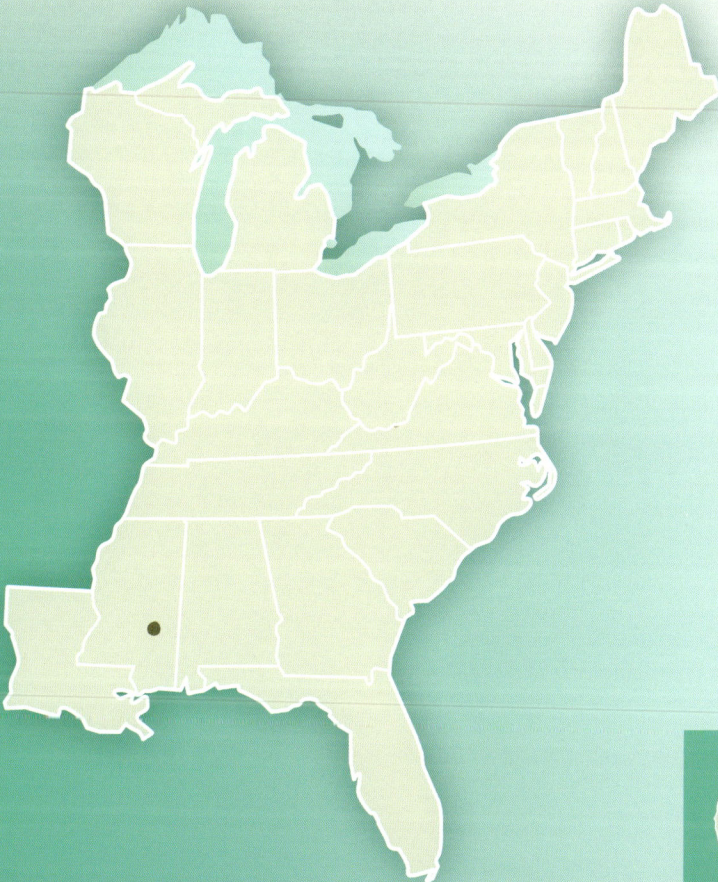

Similar Species The Bay Springs salamander resembles other *Plethodon* species
within the range but is more slender.

Distribution Known only from a location 2 miles south of Bay Springs in Jas-
per County, Mississippi.

Habitat The two known specimens
were collected in a mixed hardwood habitat with small springs and seepage areas.

Conservation No specific conservation plans are likely to be forthcoming
because the species is presumed extinct.

DID YOU KNOW?

*Some of the salamanders
discovered in the mountains
of Central America in the
late 1900s and early 2000s
became extinct because of
habitat destruction before
the scientific publications
describing them even ap-
peared in print.*

Taxonomy The Bay Springs salamander was described in 1998 from two specimens collected and preserved in 1964.

TELLICO SALAMANDER *Plethodon aureolus*

BODY SHAPE	NUMBER OF COSTAL GROOVES	PATTERN	BODY SIZE
Slender	16	Brass-colored flecks on back, and light spotting on sides	Adults up to 5.8 inches in total length

BODY COLOR
Black

Description This small (usually between 4 and 6 inches) black salamander has brass-colored markings on the back and white or yellow spots on the sides. The chin is lighter, and a mental gland is present. The belly is gray and unmarked. The light markings on the body are reduced in some individuals found at higher elevations.

Larvae and Juveniles Juveniles resemble adults.

The chin of the Tellico salamander is lighter than the dark back and tail.

top Brass-colored markings on the back and light-colored spots on the sides are characteristic of most Tellico salamanders, which are smaller than adults of most other plethodontid species found within their geographic range.

TELLICO SALAMANDER *Plethodon aureolus*

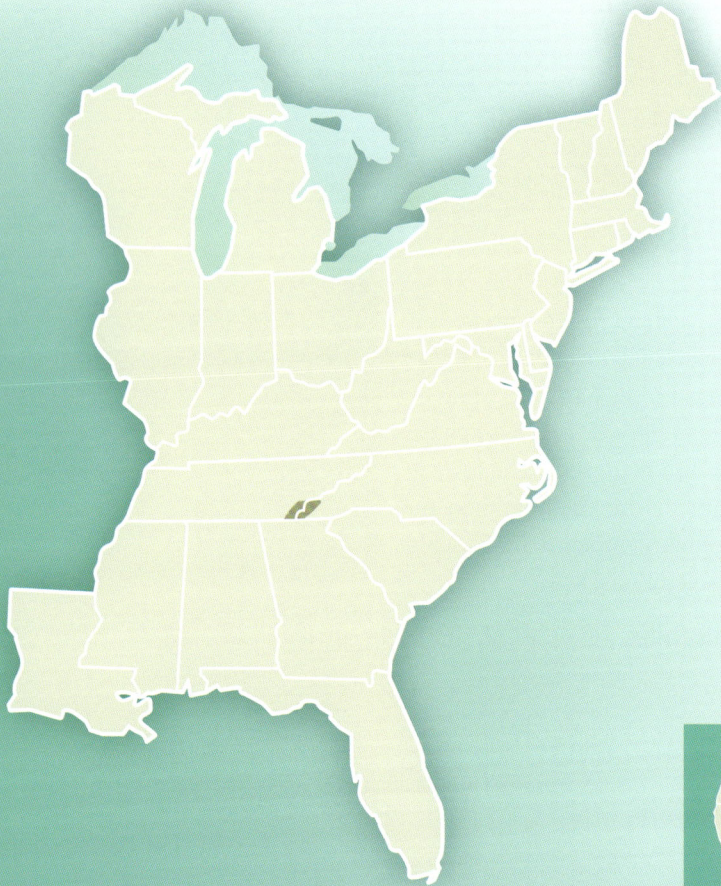

Similar Species Because of its limited geographic range and small body size, the Tellico salamander is likely to be confused only with the few *Plethodon* species with which it overlaps, most of which are larger. The brass-colored markings are the key character for identification.

Distribution The Tellico salamander is known from two counties (Graham and Cherokee) in North Carolina and two (Polk and Monroe) in Tennessee.

Habitat The species is associated mostly with hardwood forests in lowland areas but is also found at higher elevations above 5,000 feet.

Behavior and Activity Little is known of the general behavior and activity of Tellico salamanders, but they presumably spend most of their time beneath logs, dead leaves, and other ground litter or deeper underground.

Reproduction Tellico salamanders have an elaborate terrestrial courtship dance. Nest sites and clutch sizes are unknown, but presumably the eggs are laid underground. Embryos develop in the egg and hatch as miniatures of the adult form, and juveniles are seen above ground in August.

Food and Feeding They presumably eat small terrestrial invertebrates.

Predators and Defense Probable predators are small snakes, large spiders, shrews, and birds.

Conservation No serious conservation threats have been identified for this species.

Taxonomy The Tellico salamander was described in 1984 and has no subspecies. Tellico salamanders hybridize with red-cheeked salamanders.

PIGEON MOUNTAIN SALAMANDER *Plethodon petraeus*

BODY SHAPE
Flattened and slender; tail more than half the total length

NUMBER OF COSTAL GROOVES
15 or 16

BODY COLOR
Brown and black

PATTERN
Irregular brown band down back; black tail and sides with white or brassy spots

BODY SIZE
Adults up to 7.2 inches in total length

Description These are large salamanders, and females are slightly larger than males. The basic body color of the back and belly is black, but a broad, irregular brown (reddish to olive) band or patch extends the length of the back and onto the sides. The back, sides, limbs, and tail have white or brass-colored spots. Expanded toe tips and an elongated toe on each foot help these salamanders climb rock faces. Males have a noticeable round mental gland.

Larvae and Juveniles Young juveniles have brownish or brass-colored spots on the back that eventually connect to form the pattern characteristic of adults.

top The Pigeon Mountain salamander is found only in Walker and Chattooga counties in northwestern Georgia and can be distinguished from other salamanders in the region by its large size and brown back.

Although geographically rare, this salamander can be locally common under the right weather conditions.

Similar Species The brown coloration on the back and expanded toe pads readily distinguish the Pigeon Mountain salamander from other species in the region.

Distribution The total range is limited to the eastern side of Pigeon Mountain in Walker and Chattooga Counties in northwestern Georgia.

Habitat The Pigeon Mountain salamander is strongly associated with limestone elements of rocky cliffs, crevices, and cave openings. The surrounding woodland habitat is primarily oak-hickory forest.

Behavior and Activity Pigeon Mountain salamanders have physical adaptations that help them climb the sides of rock cliffs and crevices. They are active throughout the year except during the coldest winter temperatures. They can be found under rocks or in caves during most of the year, and in early spring can also be found beneath logs in wooded areas. They are most active above ground on rock surfaces and the forest floor, especially during rains. When conditions are dry, these salamanders will often keep the body concealed in a rock crevice or beneath a rock slab with only the head visible. During dry periods in summer they move into caves or deep crevices.

During dry weather, Pigeon Mountain salamanders will often keep their body concealed in a rock crevice or beneath a rock slab with only the head visible.

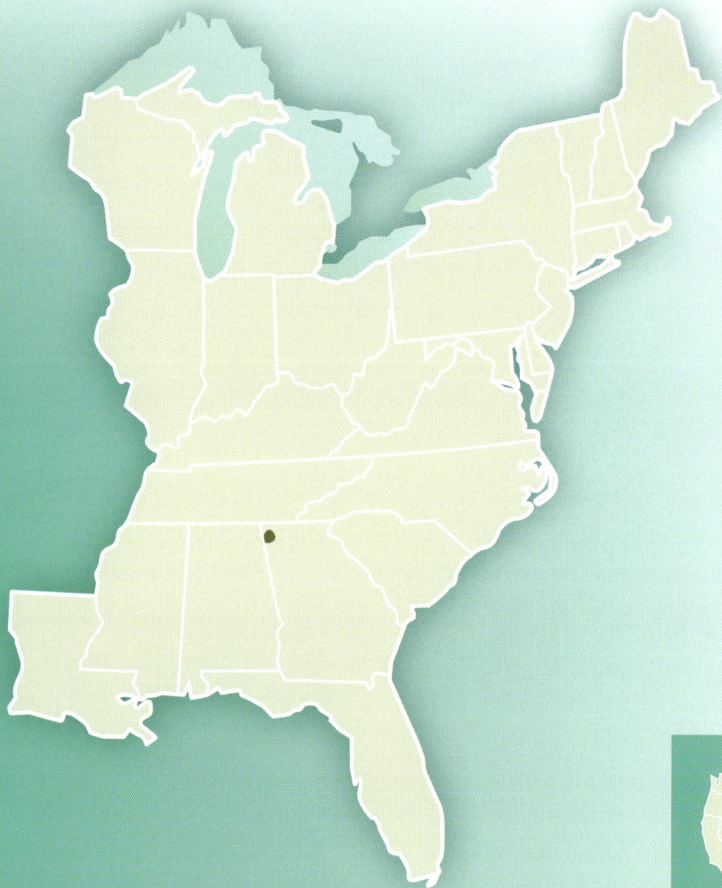

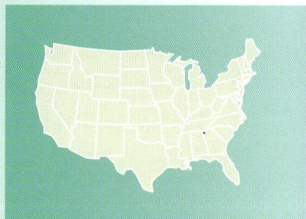

Reproduction Complete details of the reproductive cycle of the Pigeon Mountain salamander are unknown other than that the breeding season is presumed to be in spring and summer, and eggs are laid later in the summer or in the fall. Clutch sizes of around 17 have been reported. Nests in nature are unknown, but presumably the eggs are laid underground.

Food and Feeding Primary food items are small invertebrates including terrestrial insects, spiders, and worms.

The Pigeon Mountain salamander is associated with limestone topography of rocky cliffs and crevices that are surrounded by woodland habitat.

Predators and Defense Predation has not been observed, but small snakes, birds, and shrews probably prey on these salamanders.

Conservation The species has a very restricted geographic range and habitat requirements and is protected by the state of Georgia as a Rare species.

Taxonomy The Pigeon Mountain salamander was discovered and recognized as a new species in 1972 and was described in the scientific literature in 1988. No subspecies are recognized.

SOUTHERN APPALACHIAN SALAMANDER *Plethodon teyahalee*

BODY SHAPE	NUMBER OF COSTAL GROOVES	PATTERN	BODY SIZE
Large and stout	16	Small white spots on back, larger ones on sides	Adults up to 8.2 inches in total length
	BODY COLOR Black		

Description These relatively large black salamanders have tiny white spots on the back and tail, and larger white spots on the sides. Some individuals have red spots on their legs. The underside is grayish, with the chin being a lighter gray. The round mental gland is apparent in breeding males.

Larvae and Juveniles Juvenile southern Appalachian salamanders have the same color and markings as adults.

Similar Species Other woodland salamanders found within this species' geographic range and habitats do not have white spotting on the body and red spots on the legs. However, the southern Appalachian salamander hybridizes with the

top This southern Appalachian salamander can be distinguished from most other terrestrial salamanders in its range by its larger size and the presence of white spotting on the body and red pigment on the legs.

red-cheeked salamander and possibly other species of woodland salamanders in some areas, and individuals with a mix of color traits are difficult to identify.

Distribution The southern Appalachian salamander is restricted to sites in the Blue Ridge physiographic province from southwestern North Carolina to northwestern South Carolina, southeastern Tennessee, and Rabun County in northeastern Georgia. Individuals have been found at elevations approaching 5,000 feet.

Southern Appalachian salamanders are most active above the ground during rains at night.

Habitat Adults and juveniles are found in deciduous hardwood forests.

Behavior and Activity Like most other woodland salamanders, this species leads a secretive existence by remaining underground or beneath rocks, logs, dead leaves, and other ground litter most of the time. They are most prevalent above the ground at night during rains.

Reproduction Reproduction is terrestrial and presumably similar to the red-cheeked salamander.

Food and Feeding Southern Appalachian salamanders probably eat most small terrestrial invertebrates they can capture and are known to consume snails, millipedes, sowbugs, spiders, springtails, and a variety of insects including beetles and ants.

Predators and Defense No predators have been verified, but small snakes, birds, and shrews probably prey on southern Appalachian salamanders. Aside from leading a clandestine existence and avoiding some predators by coming out only at night, these salamanders produce skin secretions that are mildly toxic or offensive to some predators.

Conservation Concerns for southern Appalachian salamanders apply primarily to local populations that might suffer from degradation of their terrestrial habitat, although the species has been reported to persist in second-growth forests after timbering.

Taxonomy The southern Appalachian salamander was described in 1950 as *P. jordani teyahalee*, a subspecies of the red-cheeked salamander, and in 1983

The southern Appalachian salamander is restricted to sites in the Blue Ridge physiographic province

was recognized as a distinct species named P. *oconaluftee*. The latter name did not follow the rules of scientific nomenclature, and the species name was later changed to P. *teyahalee*. No subspecies are recognized.

RAVINE SALAMANDERS

Northern Ravine Salamander *Plethodon electromorphus*

Southern Ravine Salamander *Plethodon richmondi*

BODY SHAPE	NUMBER OF COSTAL GROOVES	PATTERN	BODY SIZE
Slender; short legs	19–23	Body and tail speckled with silvery or brassy flecks	Adults up to 5.6 inches in total length
	BODY COLOR		
	Dark brown to black		

Description The body is dark brown or black, and the head, body, and tail are speckled with silver or brass-colored flecks. The underside is similarly dark and has flecks on the chin but not on the belly. The tail constitutes about half the total length, which ranges from about 3 to 5.6 inches. Northern ravine salamanders and southern ravine salamanders are identical externally. Where their ranges overlap in northern Kentucky they can be distinguished only by genetic analysis.

top Southern ravine salamanders inhabit hardwood forests in rocky areas and are more likely to be found beneath rocks or in rock crevices than under logs or leaves. They are found near the surface most often in spring and fall.

The belly of ravine salamanders is relatively dark, and the tail is as long as the rest of the body.

Larvae and Juveniles Recently hatched juveniles are light gray and about 1 inch in total length. Older juveniles develop adult coloration.

Similar Species Several species whose ranges partially overlap that of the southern ravine salamander in the Southeast are superficially similar but can usually be differentiated by body coloration, body shape, or number of costal grooves. The northern zigzag salamander usually has reddish coloration on the belly; Weller's and Wehrle's salamanders have fewer costal grooves; slimy salamanders have a more robust body; and northern gray-cheeked salamanders have gray cheeks.

Distribution The northern ravine salamander ranges from southwestern Pennsylvania through Ohio to southeastern Indiana, and in the Southeast is found only on the northern edge of Kentucky. The southern ravine salamander occurs in southern West Virginia and in parts of four southeastern states: eastern Kentucky, western Virginia, northeastern Tennessee, and northwestern North Carolina.

Habitat Both species characteristically occur on slopes of hardwood forests with rock outcrops.

When threatened by the presence of a predator, ravine salamanders often remain motionless in an attempt to go unobserved.

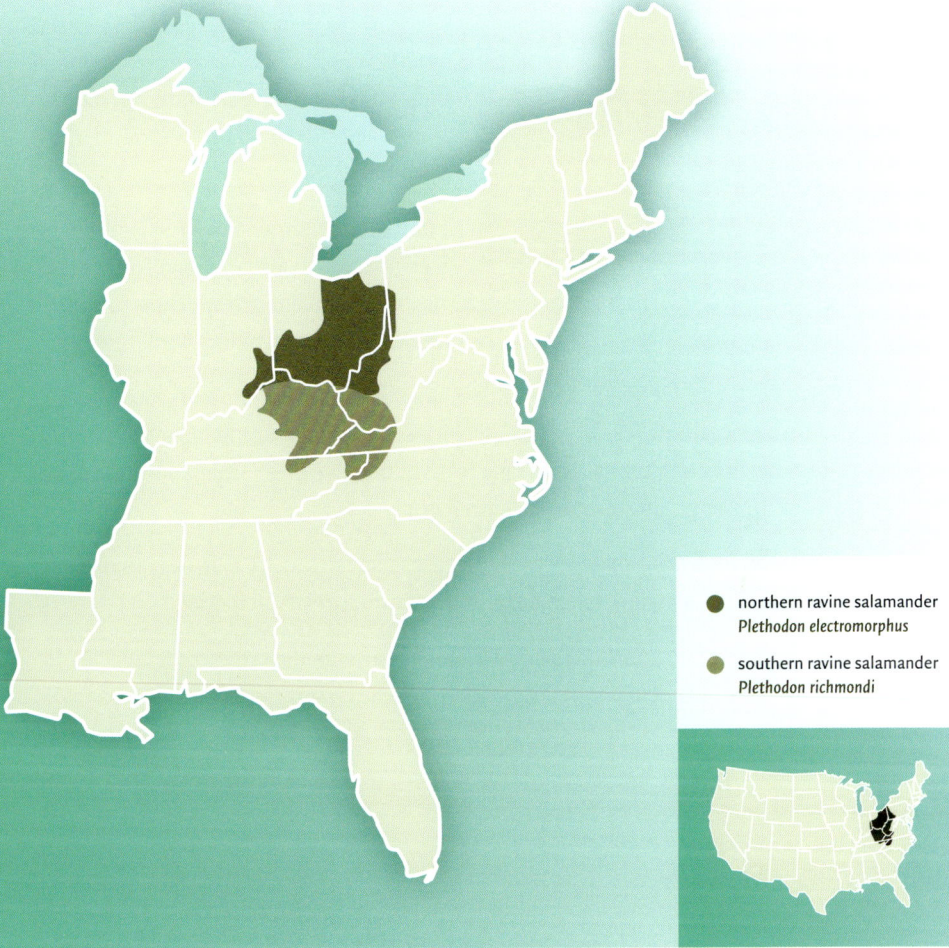

northern ravine salamander
Plethodon electromorphus

southern ravine salamander
Plethodon richmondi

Behavior and Activity Ravine salamanders are secretive and often remain below the surface in rock crevices or underground tunnels. They are more likely to be found under rocks than logs and are most likely to be near the surface during fall, winter, and spring.

Reproduction Mating takes place as early as November and as late as May; the timing depends on local environmental conditions and other unknown factors. Females lay their eggs underground or beneath rocks from April through June, with local variation among populations. Each female lays 5–15

The northern ravine salamander is indistinguishable in appearance from the southern ravine salamander and must be identified on the basis of genetic analysis and geographic location.

eggs that hatch by early fall, but hatchlings usually remain underground until spring.

Food and Feeding Ravine salamanders feed on a wide variety of small terrestrial invertebrates in all months of the year. Among the major groups eaten are ants, sowbugs, worms, beetles, snails, centipedes, roaches, spiders, and mites.

Predators and Defense Ringneck snakes commonly eat ravine salamanders, and other small snakes, large spiders, birds, and small mammals such as shrews probably prey on them as well. One form of defense is to remain motionless in an attempt to go unobserved by a predator.

Conservation The primary threat to these species is loss or degradation of habitat resulting from urban development and forestry activities.

Taxonomy The southern ravine salamander was described as a species in 1938. The northern ravine salamander, which was originally considered to be the same species as the southern, was described as a separate species in 1999. Neither is partitioned into subspecies.

WEBSTER'S SALAMANDER   *Plethodon websteri*

BODY SHAPE
Small and slender

**NUMBER
OF COSTAL
GROOVES**
17–19

BODY COLOR
Brown or reddish
brown

PATTERN
Usually a broad
stripe down the back
and onto the tail

BODY SIZE
Adults up to 3.3
inches in total
length

Description Webster's salamanders are tiny brown salamanders that range in color from reddish to yellowish brown. Some individuals have a stripe down the back to the end of the tail, and a small proportion of the population has no stripe at all. The sides of the stripe are wavy and are usually brighter on the front part of the tail than on the body. The sides have lighter-colored flecks, and the belly is mottled with orange, red, black, and white.

Larvae and Juveniles Recent hatchlings are about 0.5 inches long and resemble adults in coloration.

Similar Species Webster's salamanders can be confused with zigzag and southern red-backed salamanders, but the ranges of the three species overlap

top Webster's salamanders inhabit rocky areas in hardwood forests with heavy ground litter.

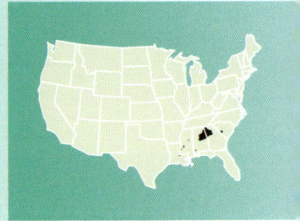

in only a few areas. Because these species are variable in color pattern, all three must be distinguished by genetic analysis.

Distribution Webster's salamanders are found from South Carolina to Louisiana, but except for a relatively broad area from western Georgia through east-central Alabama they occur as isolated populations.

Habitat Hardwood forests in rocky terrain with heavy ground litter of leaves, logs, and rocks are the preferred habitat.

Behavior and Activity These salamanders are locally abundant on the surface beneath ground litter from October to May but virtually disappear deep underground in summer through early fall.

Reproduction Terrestrial courtship and mating generally occur from January to late March, but Webster's salamanders can be found mating as early as late November in southern Mississippi. Females begin laying their clutch of 3–8 eggs underground in June and July. Hatchlings are found above ground in October.

Food and Feeding Webster's salamanders eat small terrestrial invertebrates such as ants, mites, termites, springtails, spiders, snails, centipedes, flies, and worms.

Predators and Defense Specific predators have not been documented but presumably include ringneck snakes, garter snakes, large spiders, and shrews that occupy similar habitat. Observations of populations in southwestern Mississippi provide reliable evidence that predation by turkeys has a significant impact on Webster's salamanders, especially when they are migrating to breeding outcrops. Often the salamanders have missing tails, perhaps due to turkeys targeting the brighter tail than the body.

Conservation No generalized conservation threats have been identified for Webster's salamander other than loss of forested habitat. South Carolina lists the species as Endangered, and Louisiana considers it a Species of Special Concern.

Taxonomy Populations of Webster's salamanders were reported in the scientific literature as early as the 1950s, but as zigzag salamanders. The species was described in 1979, and no subspecies have been recognized.

> **DID YOU KNOW?**
>
> New species of salamanders continue to be described in the eastern United States as DNA analyses differentiate species that look identical. Several new species are expected to be described within the next several years, even as the actual number of salamander populations continues to decline.

CHEOAH BALD SALAMANDER *Plethodon cheoah*

BODY SHAPE
Slender

NUMBER OF COSTAL GROOVES
16

BODY COLOR
Uniformly blue-black; usually some red on legs

PATTERN
White to yellow spots, primarily on the sides

BODY SIZE
Adults up to 5 inches in total length

Description The body of these medium-sized salamanders is uniformly blue-black except for white to yellow spots on the sides. Usually less than half of the upper surface of the legs is covered by red pigment. The belly is light gray with or without light yellow spots. About half the individuals in any population have a small amount of red pigment on their cheeks.

Larvae and Juveniles Young ones less than 1.25 inches in total length have red spots and brassy flecking on the back and may have a red V atop the head with the apex pointing toward the snout.

Similar Species Red-cheeked salamanders do not have red pigment on the upper surface of the legs but are otherwise difficult to distinguish from Cheoah

top and opposite Most Cheoah Bald salamanders have red pigment on part of their legs, but only about half of the individuals have any red pigment on their cheeks.

Bald salamanders. Confirmation of identification in the zone where the two species overlap requires molecular analysis.

Distribution The species is restricted to Cheoah Bald in Graham and Swain Counties in North Carolina and ranges to elevations above 5,000 feet.

Habitat Higher-elevation forests in rough, rocky terrain are the preferred habitat.

Behavior and Activity Little is known of the year-round activity patterns of the Cheoah Bald salamander other than that individuals move underground during the coldest and driest periods. They usually disappear during the hot, dry days of summer, but can be quite common during late spring and very early summer. When on the surface, they stay under rocks and logs.

Reproduction Following courtship and mating on the ground surface, females are presumed to retreat underground to lay their eggs, and to stay with the eggs until they hatch. Clutch size and incubation period are unknown.

Food and Feeding Cheoah Bald salamanders are presumed to eat an array of terrestrial invertebrates.

Predators and Defense Probable predators are small snakes, large spiders, and shrews.

Conservation No general conservation threats are imminent for this species, which is not protected in its limited geographic range in North Carolina.

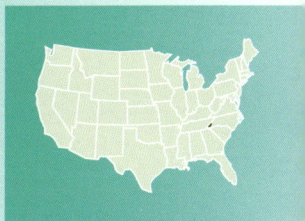

Taxonomy Originally presumed to be a local population of red-cheeked salamander, the Cheoah Bald salamander was described as a new species in 2000 and has no recognized subspecies. Cheoah Bald salamanders hybridize with southern Appalachian salamanders.

Cheoah Bald salamanders have a limited geographic range, restricted to two counties in North Carolina. The species can be found at elevations above 5,000 feet.

ZIGZAG SALAMANDERS

Northern Zigzag Salamander	*Plethodon dorsalis*
Southern Zigzag Salamander	*Plethodon ventralis*

BODY SHAPE
Small and stout

NUMBER OF COSTAL GROOVES
17–19, usually 18

BODY COLOR
Brown or dark gray

PATTERN
Reddish to yellowish stripe on back and tail, or no stripe

BODY SIZE
Adults up to 3.5 inches in total length

Description Zigzag salamanders are small, brownish or dark gray salamanders that vary considerably in color and pattern. The color ranges from reddish to yellowish brown, and individuals either have a wavy central stripe down the back and onto the tail or no stripe at all. The sides of the stripe are scalloped toward the front of the body. The dark sides have light-colored flecks. The belly is a mix of orange, red, black, and white. A pale red shoulder patch is present on most individuals in middle Tennessee.

top Northern zigzag salamanders look exactly like southern zigzag salamanders, and some salamander experts consider them to be the same species.

Larvae and Juveniles Young juveniles about 2 inches long appear on the surface in September. Although this terrestrial salamander bypasses the aquatic larval stage and develops completely within the egg, remnant gill stubs are occasionally present in hatchlings. Juveniles resemble adults. The two zigzag species are indistinguishable in appearance but can usually be identified by geographic location.

Similar Species The southern zigzag salamander's geographic range overlaps that of Webster's salamander only in Alabama (Bibb and Jefferson Counties). The southern red-backed salamander looks similar but has more costal grooves (18–22).

Distribution The inclusive geographic range of the northern and southern zigzag salamanders is from southern Illinois and Indiana south through Kentucky to Mississippi; the two species overlap in eastern Kentucky. The southern zigzag salamander is restricted to the Southeast, occurring in locations in western Virginia and North Carolina, eastern Kentucky and Tennessee, northwestern Geor-

Zigzag salamanders bypass the aquatic larval stage and develop completely within the egg in the terrestrial habitat.

gia, northern Alabama, and the extreme northeastern corner of Mississippi. A few populations in northern Alabama right on the border of Tennessee may be northern zigzag salamanders.

Habitat The primary habitat of zigzag salamanders is hardwood forests, often in rocky areas. During the warm months they are common inhabitants

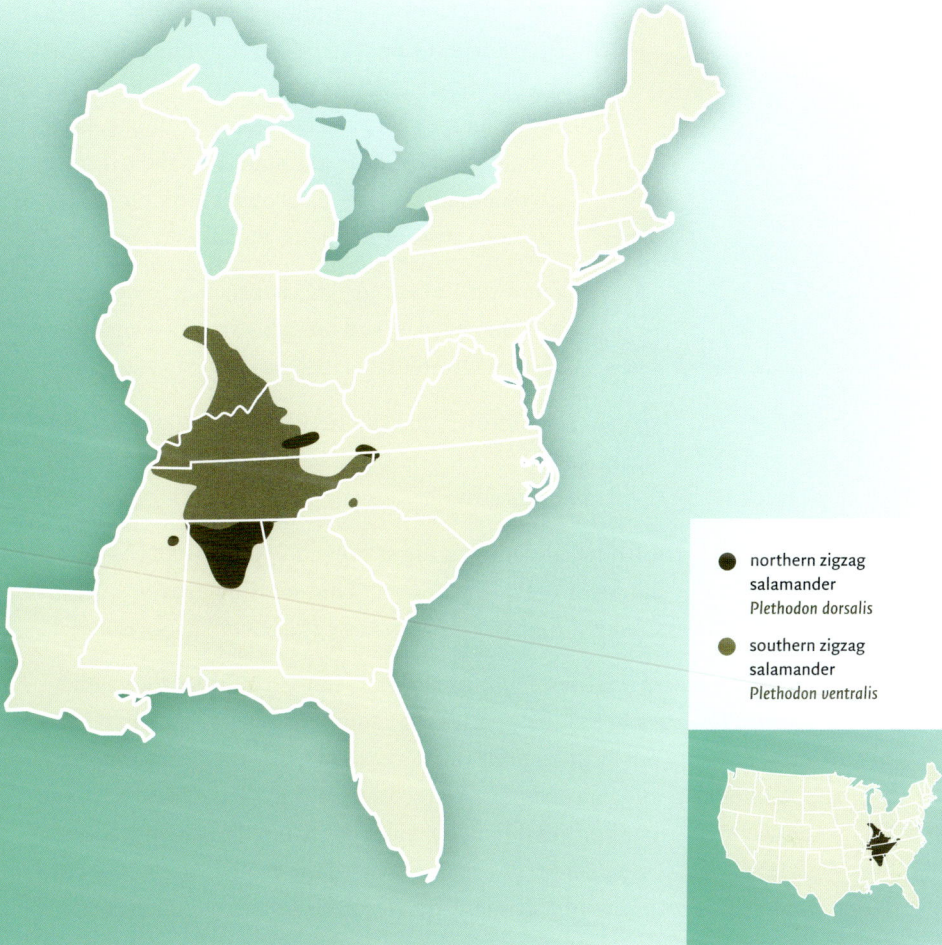

northern zigzag
salamander
Plethodon dorsalis

southern zigzag
salamander
Plethodon ventralis

of subterranean passages and can often be found in crevices in cave floors
and walls.

Behavior and Activity   During fall and winter, zigzag salamanders commonly
stay beneath the ground litter of dead leaves, logs, and rocks during the day and
walk around on the ground, roots, bases of shrubs, and rocks while foraging
at night during wet weather. With the onset of warm, dry weather they retreat
underground or into caves, where they remain until the cooler and damper
conditions of fall return.

A female zigzag salamander lays up to 10 eggs and stays with them during development, which may take 3 months.

Reproduction Courtship and mating take place from fall to spring, but the specific timing of reproduction varies among populations on the basis of winter temperature and moisture conditions. Each female lays 1–10 eggs and stays with them during development, which takes about 3 months. Very few nests have been found, and all were in caves.

Food and Feeding Zigzag salamanders eat spiders and beetles, and probably any other small terrestrial invertebrates they can catch and swallow.

Predators and Defense Screech-owls and ringneck snakes are known predators, and garter snakes, large spiders, and shrews and other small mammals probably eat zigzag salamanders as well.

Conservation Local populations of zigzag salamanders can be reduced in size or eliminated by improper timbering or other land-clearing activities. North Carolina lists the southern zigzag salamander as a Species of Special Concern.

Taxonomy The original description of P. dorsalis in 1889 included the southern and northern forms as a single species. The southern zigzag salamander was described in 1997 as a distinct species on the basis of genetic evidence. Northern and southern zigzag salamanders interbreed naturally in parts of Kentucky, and some salamander authorities do not consider the two to be separate species. The Ozark zigzag salamander (P. angusticlavius) of the Ozark Plateau was once considered a subspecies of P. dorsalis, and Webster's salamander was included in the zigzag salamander species until 1979.

RED-LEGGED SALAMANDER *Plethodon shermani*

BODY SHAPE	NUMBER OF COSTAL GROOVES	PATTERN	BODY SIZE
Large and slender; protruding eyes	15 or 16	Solid back with red blotches on legs	Adults up to 6.7 inches in total length
	BODY COLOR Bluish black		

Description These large salamanders have a gunmetal blue–black back and tail. The belly is bluish gray. The bright red markings on the legs are a signature trait of the species, but red coloration is reduced or absent on some individuals. Females get larger than males.

Larvae and Juveniles By the time they appear on the surface, juveniles resemble adults.

Similar Species Red-legged salamanders hybridize with southern Appalachian salamanders at several locations, and genetic analysis may be required to determine the two species.

top Red-legged salamanders have a dark blue or black back, typically with bright red markings on the legs.

Although red legs are characteristic of most red-legged salamanders, some individuals lack the red coloration and have black legs.

Distribution The species is found in southwestern North Carolina in the Nantahala Mountains and in Towns County in northwestern Georgia, and has been reported from Monroe County in southeastern Tennessee.

Habitat Red-legged salamanders occupy wet, cool mountain hardwood forests with rhododendron.

Behavior and Activity These woodland salamanders remain beneath logs, fallen bark, and rocks during the day and may emerge onto the surface at night or during rains. During cold winters they retreat deeper underground.

Reproduction Courtship and mating are terrestrial, and eggs are presumably laid underground. Clutch size and incubation period are unknown.

Food and Feeding Red-legged salamanders eat a wide array of terrestrial invertebrates including many insects (ants, beetles, flies, wasps, aphids, and craneflies), spiders, springtails, millipedes, and earthworms.

Predators and Defense Predators presumably include small terrestrial snakes, large salamanders, shrews, and birds, but nothing has been published on the subject.

Conservation Despite the relatively small geographic range, the habitat of red-legged salamanders is mostly in national forests and does not appear to be threatened by development or timbering activities.

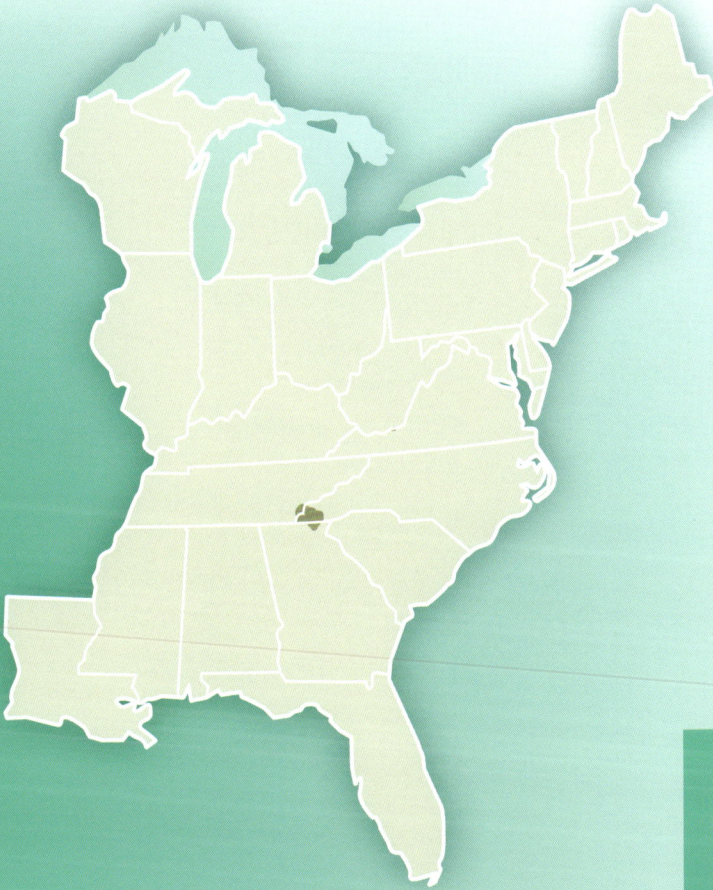

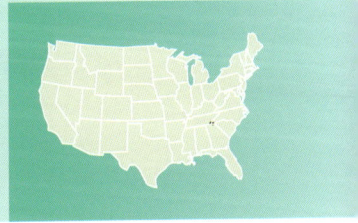

Taxonomy The red-legged salamander was described as a distinct species in 1906. Some herpetologists later classified it as a subspecies of slimy salamander before it was placed in the *Plethodon jordani* complex as a subspecies. It was reinstated as a full species in 2000.

RED-CHEEKED SALAMANDER *Plethodon jordani*

BODY SHAPE
Large and slender;
protruding eyes

NUMBER OF COSTAL GROOVES
15 or 16

BODY COLOR
Black

PATTERN
Solid back with
reddish cheeks

BODY SIZE
Adults to 6.7 inches
in total length

Description These large salamanders have a black back and tail, bluish gray belly, and pink or red cheeks. The head is sometimes brownish. The red coloration is sometimes reduced or, rarely, absent entirely.

Larvae and Juveniles Juveniles remain underground for almost a year after hatching and appear on the forest floor in May. They resemble the adults.

Similar Species The red cheeks and black body readily distinguish the red-cheeked salamander from most other salamanders within its geographic range. Red-cheeked salamanders can hybridize with southern Appalachian salamanders, and differentiating between the two species requires genetic analysis.

top The back and tail, bluish gray belly, and distinctive pink or red cheeks are trademarks of the red-cheeked salamander, although the red coloration is absent in some specimens.

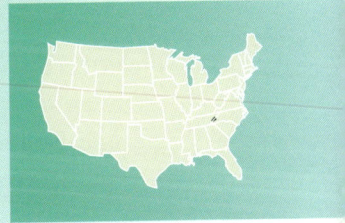

Distribution Red-cheeked salamanders are found in Tennessee and North Carolina within the boundaries of Great Smoky Mountains National Park.

Habitat Slopes and ridges with hardwoods and spruce-fir forests with abundant ground cover of mosses, logs, and rocks are preferred habitats.

Behavior and Activity These salamanders remain hidden beneath logs, moss, and rocks during the day and crawl around on the surface at night and during rains. They retreat deeper underground when the surface is dry.

Reproduction All aspects of reproduction are terrestrial. Clutch size is unknown; eggs are presumably laid in underground burrows.

Food and Feeding Numerous invertebrates have been recorded in the diet of red-cheeked salamanders, with expected variations from season to season and among geographic locations. Their diet includes millipedes, insect larvae, worms, snails, beetles, spiders, springtails, flies, and ants.

Predators and Defense Black-bellied salamanders, spring salamanders, eastern garter snakes, and short-tailed shrews eat red-cheeked salamanders. They are probably also preyed on by ringneck snakes and birds, although their noxious skin secretions may repel some predators.

Conservation Red-cheeked salamanders occur within the Great Smoky Mountains National Park and are not threatened from a conservation perspective. However, global warming could be a future threat because this species depends on cool, moist conditions.

Taxonomy The red-cheeked salamander was described as a species in 1901. It is sometimes known as Jordan's salamander and may be called the red-necked salamander in older field guides.

COW KNOB SALAMANDER *Plethodon punctatus*

BODY SHAPE	NUMBER OF COSTAL GROOVES	PATTERN	BODY SIZE
Slender	17 or 18	Body, legs, and tail with white or cream-colored spots	Adults up to 6.2 inches in total length
BODY COLOR Dark gray to black			

Description Cow Knob salamanders are large and grayish black with a profusion of irregular white to cream-colored spots on the body, legs, and tail; the spots are more abundant on the sides than on the back. The throat is light to pinkish gray, and the belly is dark gray with small cream-colored spots. The Cow Knob salamander superficially resembles a slimy salamander but has thinner, more delicate-looking legs. Its head is flatter as well, and the eyes are more protruding. The top of the head is often slightly purplish relative to the gray-black body.

Larvae and Juveniles Juvenile Cow Knob salamanders appear on the surface in September and are similar to the adults in appearance, although they have fewer light spots.

top Cow Knob salamanders are found in mixed hardwood forests at high elevations, most commonly on cool, north-facing slopes.

Similar Species Cow Knob salamanders superficially resemble two species of slimy salamanders that occur in the area. Northern slimy salamanders are patterned with brassy to silvery spots. White-spotted slimy salamanders have large, irregular white spots. Both have 16 or fewer costal grooves. The Cow Knob salamander is closely related to Wehrle's salamander, but the two species apparently do not overlap geographically and although similar in appearance, the Cow Knob salamander lacks the red pigment and brassy flecking of Wehrle's salamander.

Distribution Cow Knob salamanders occur on Shenandoah Mountain and Great North Mountain in the George Washington National Forest in Augusta, Rockingham, and Shenandoah Counties in Virginia; and Pendleton and Hardy Counties in West Virginia.

Habitat This species is restricted to mixed hardwood forests in rocky areas at higher elevations (above 2,500 feet), and most populations occupy cooler, north-facing slopes. These salamanders are most common in areas of weathered talus rock where soil and leaf litter are also plentiful. The largest populations have been found in association with old hemlock forest stands.

Behavior and Activity Cow Knob salamanders are generally active above ground in May and June, and again—but not as abundantly—in September and October when environmental conditions are wet and cool. They remain under logs and rock cover during the day but are active on the surface at night as they forage on the ground, rocks, and the base of trees. They retreat underground in July and August when surface conditions are dry.

Reproduction Cow Knob salamanders are presumed to breed terrestrially and to lay eggs underground in the summer. No nests have been found or described.

Cow Knob salamanders are grayish black with a profusion of irregular white to cream-colored spots on the body, legs, and tail.

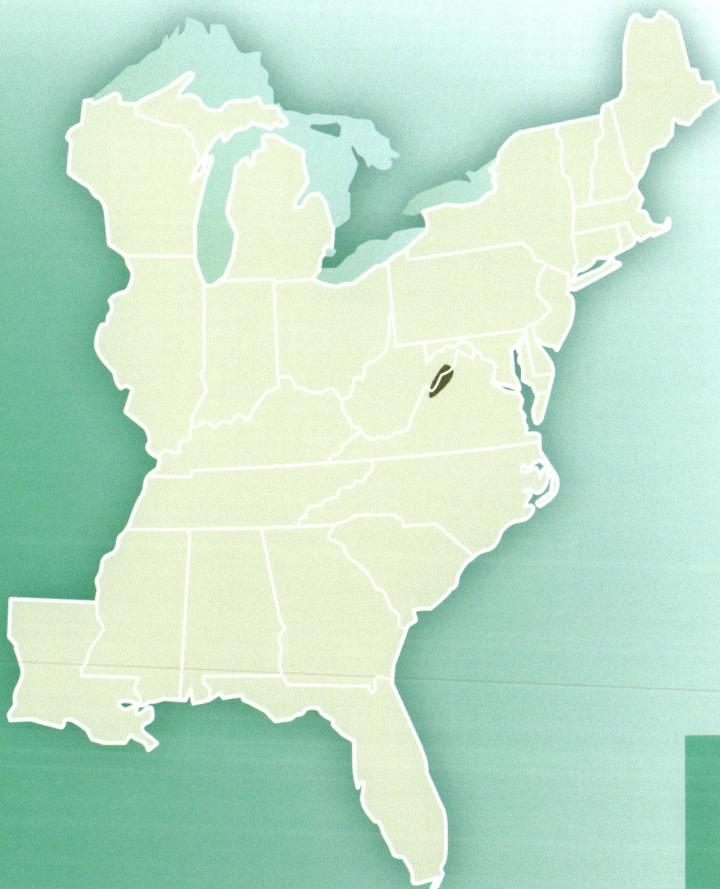

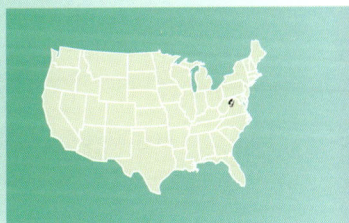

Each female lays 7–16 eggs and probably remains with them until hatching, as is typical of most members of the genus.

Food and Feeding Cow Knob salamanders eat a variety of small invertebrates, including ants, springtails, beetles, flies, millipedes, centipedes, spiders, and mites. Adults consume larger and more varied prey than juveniles.

Predators and Defense Small snakes, birds, shrews, ground-feeding birds, and possibly raccoons and opossums are probably this salamander's primary predators, although no specific observations have been published. Toxic secre-

The very restricted geographic range of the Cow Knob salamander makes the species potentially vulnerable to timber operations.

tions produced by glands in the tail are presumably effective at discouraging some predators.

Conservation The very restricted geographic range of the Cow Knob salamander makes the species potentially vulnerable to timber operations. The regional decline in hemlock trees because of the introduced hemlock woolly adelgid, loss of the canopy because of the gypsy moth, and climate change threaten the habitat. The U.S. Fish and Wildlife Service lists the Cow Knob salamander as a Species at Risk, and both Virginia and West Virginia list it as a Species of Special Concern. This means that prior to implementation all management activities in the Shenandoah Mountain Crest Special Biological Area in the George Washington National Forest, where much of its range lies, must be reviewed for their potential impact on the salamander.

Taxonomy The Cow Knob salamander was officially described as a species in 1972. No subspecies are recognized. This species has also been called the white-spotted salamander.

people and salamanders

WHAT IS A HERPETOLOGIST?

A scientist who studies reptiles and amphibians (including salamanders) is called a herpetologist. Many zookeepers, pet hobbyists, and wildlife managers also study reptiles and amphibians, often with interests in specific groups.

Why Do Herpetologists Study Salamanders? Salamanders can be studied in classrooms, in laboratories, and in the wild. One of the first animals pre-medical students have traditionally studied in a comparative anatomy course is a large salamander—the mudpuppy (*Necturus maculosus*). Students dissect pre-served mudpuppy specimens to examine the circulatory, nervous, respiratory, muscular, and skeletal systems, which have the basic design of these systems in humans. Biological supply houses have provided hundreds of thousands of mudpuppy specimens for anatomy courses.

Salamander species that are easily maintained in captivity have been used extensively in laboratory experiments to address questions related to how sala-manders defend themselves from predators, whether they engage in competitive behavior toward other salamanders, and whether they can recognize relatives as individuals. Larval and adult tiger salamanders have been used for decades to address developmental biology questions related to regeneration of body parts.

Some conservation biologists fo-cus their research on salamanders

above A researcher from Virginia Tech releases a protected hellbender into an Appalachian river in southwestern Virginia.

left These marbled salamanders were collected overnight in a single bucket at a drift fence by researchers at the Savannah River Ecology Labo-ratory in Aiken, South Carolina.

above The Shenandoah salamander (*left*) and red-backed salamander (*right*), both members of the genus *Plethodon*, can be difficult to distinguish. Note the differing widths of the dorsal stripe.

left Although in different genera, the red-cheeked salamander (*top*) and imitator salamander (*bottom*) can be difficult to tell apart based on appearance.

because many species in the United States and elsewhere are currently at risk. Salamander populations that are sensitive to forestry or mining activities or to urban development are good indicator species; that is, their status indicates the health of the overall ecosystem.

How Do Herpetologists Study Salamanders? Herpetologists who study salamander ecology and behavior in the field use many techniques; some are standard for all studies, and some are designed for specific situations. The basic biological information gained from such studies is often applicable to issues related to conservation of particular salamander species or habitats. Among the typical measurements herpetologists record in field studies of salamanders are body, or snout–vent, length (the distance from the tip of the snout to the vent—the posterior margin of the cloacal opening); tail length; and body weight. Date, time, and location of capture are routinely recorded too, as is the sex of each individual when it can be determined. If genetic studies are being conducted on a salamander population, tissue samples of blood or muscle (such as a toe or the tip of the tail) may be taken for DNA analyses.

Many herpetological field studies use the mark-release-recapture method to assess population size, determine individual growth rates, and estimate survivorship patterns. Salamanders can be marked for individual identification by clipping toes according to a coding scheme that gives unique combinations. This widespread, although somewhat controversial, technique has proven very

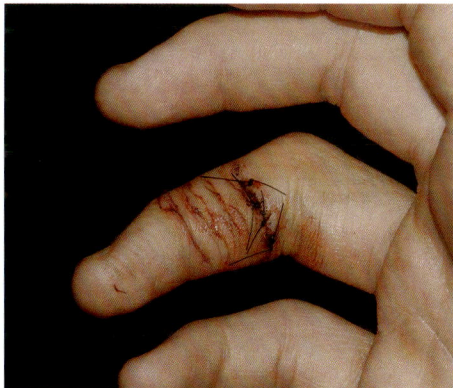

effective in long-term studies of salamanders. The elastomer technique is a more modern method that has also been used with success in field studies. An elastomer is a silicone-based, fluorescent material that can be injected beneath the skin of a salamander and seen with an ultraviolet light source. Known as the "visible implant elastomer" (VIE) tagging system, the technique uses different colors implanted at different points on the body, giving hundreds of possible combinations so that individuals can be recognized. Elastomer implants are presumed to last for the life of the marked animal and are therefore useful in long-term mark-release-recapture studies. Digital photography allows individual salamanders to be identified using body color patterns, which are permanent in adult marbled salamanders and newts, and

All salamanders have teeth, and some of the larger species can give a painful bite (hellbender, *left*; amphiuma, *right*). Such injuries are rare, however, and nearly always occur after someone picks up the animal.

presumably in other species as well. This approach is similar to analyzing a fingerprint record. Another approach for individually marking animals that has had success with larger salamanders is the use of passive integrated transponder (PIT) tags—tiny, glass-encapsulated electronic devices that can be injected into the body cavity of the salamander. A special hand-held reader determines the specific code emitted by the PIT tag.

Several techniques are used to collect salamanders—both aquatic and terrestrial forms—in the wild. If the season is right, the fundamental approach of turning over logs, rocks, and other ground litter is effective for finding the adult stage of most species. Herpetologists capitalize on salamanders' tendency to hide beneath ground cover by placing sheets of plywood known as coverboards in areas where terrestrial salamanders are known to occur. The coverboard technique is highly effective for collecting some species of ground-dwelling salamanders.

Vertical drift fences (which resemble the silt fences used in construction areas) with pitfall traps alongside them are a highly successful method for capturing seasonal pond–breeding salamanders. A salamander that encounters a drift fence will generally turn left or right and follow the fence, and will eventually fall into a pitfall—a bucket sunk into the ground next to the fence. A common and effective practice is to encircle an aquatic breeding site so that individual salamanders moving from the terrestrial habitat will encounter the

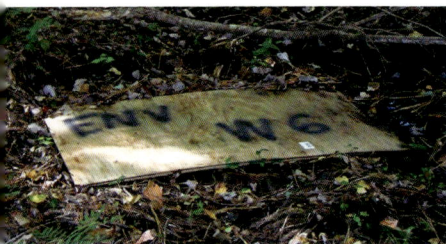

above Using coverboards is a common technique for capturing salamanders on land. These pieces of plywood placed on the ground capitalize on salamanders' tendency to hide beneath ground cover.

right Vertical drift fences with pitfall traps alongside them are a highly effective means of capturing salamanders that breed in seasonal ponds.

Several techniques are used to collect salamanders in the wild, including small funnel traps commonly used to capture minnows (top) and active dip netting (bottom) in the water.

A good way to examine aquatic salamander larvae on a field trip is to place the animals in a small plastic bag filled with water.

fence. Literally thousands of marbled salamanders (*Ambystoma opacum*) and mole salamanders (*Ambystoma talpoideum*) have been captured in a single night at the Savannah River Ecology Laboratory in South Carolina with this technique. Such exceptional capture records can occur in the fall when adults come in to breed or in the spring and summer when recently metamorphosed larvae are emigrating from the wetland.

Radiotransmitters that can be set at a particular frequency and tracked with an antenna and a receiver have been used in many wildlife studies. The development of transmitter units small enough to be implanted internally has been of great value in determining the distances individual salamanders travel into the surrounding uplands from wetlands. Pond-breeding tiger salamanders, for example, have been followed for weeks.

The study technique selected will depend on several factors, including the goal of the study, the size of the animal, and the impact a particular technique will have on the habitat and the other animals that share it. Invasive techniques are generally used only when noninvasive ones will not suffice.

A herpetologist who wants to catch salamanders for research or for a classroom project must be aware of and must follow all of the state and federal laws that have been passed to protect native wildlife. A researcher in any eastern state must obtain a scientific collecting permit in order to study salamanders in the wild regardless of whether the animals will be taken to the laboratory or simply marked and measured in the field. For example, a permit from the U.S. Fish and Wildlife Service is required for catching and measuring either species of flatwoods salamanders, which are protected under the U.S. Endangered Species Act. Collecting protected species for captive research purposes is not permitted, although if the decline continues, captive breeding programs may be necessary in the future.

Each eastern state has wildlife regulations that differ in a variety of ways, including which species are protected, how many a person can possess without a permit, and which species cannot be brought into the state or sold there. Many states also have regulations that limit or even prevent the release of salaman-

above Salamanders can usually be found under rocks in or along small streams in the Appalachians. They cannot be removed from state or national parks without a permit.

left Salamanders make excellent photographic subjects and are popular among both amateur and professional photographers.

ders that have been held in captivity. Although obtaining permits may require time-consuming paperwork and may even involve purchasing a scientific collecting permit, most state and federal regulations designed to protect native nongame wildlife such as salamanders are meant to preserve the well-being of all the species in the region.

KEEPING SALAMANDERS AS PETS

Salamanders do not have the widespread popularity as pets that many reptiles or even some frogs do. A few species, such as the fire-bellied newts (genus *Cynops*) and tiger salamanders (*Ambystoma tigrinum*), can be found commonly in the commercial pet trade, and other species sometimes become available. In fact, practically any species of salamander can be raised in captivity as long as proper housing, moisture, and food are provided. As is the case with all native wildlife species, state and federal regulations must be considered before taking a salamander from the wild because some species are protected. A person who plans to keep any wild animal in captivity must also consider ethical and conservation-related issues.

Practical and Ethical Considerations Most people who decide they want a salamander as a pet will probably want to catch their own. Before doing that, they must consider how the removal of individuals from the wild might affect the natural populations and also whether the removed individuals will do well in captivity. Opinions among herpetologists vary considerably about the capture and removal of salamanders from the wild. Some think that removing one or a few salamanders (or other herpetofauna that are not threatened or endangered) as pets will have no long-term effect on a local population. In this view, the negative environmental consequences are minimal or nonexistent, and caring for the animals and learning their habits will enhance the pet owner's respect for both the animals and their environment. Other herpetologists believe that no wild salamander should ever be taken as a pet.

Removing any animal from the wild to keep as a pet carries responsibility along with enjoyment. The pet owner becomes responsible for the animal's well-being for the rest of its life, which in some salamanders may be more than 10 or even 20 years. Someone wanting a salamander for a pet must therefore consider what would happen if it should become impossible to keep the animal. It is illegal in some states to release a pet animal back into the wild after it has been kept in captivity. If such release is legal, other questions arise concerning how local salamanders could be affected by a released pet entering the population and how the individual pet itself will fare in the wild. Will a salamander fed as a pet know how to find its own food? Does a pet salamander have a communicable disease that has not expressed itself in captivity but might be transmitted to salamanders in a wild population? Is the area where the pet is released appropriate habitat, and is the animal being released at the proper season? Is there the chance that the individual could hybridize with local salamanders of a different but closely related species and interfere with the local population dynamics and genetics? All of these issues need to be addressed by anyone who decides to have a salamander as a pet.

Salamander Husbandry Salamanders are not widely promoted in the commercial pet trade, and the life history and ecology are poorly known for most species. Permanently aquatic species such as sirens, mudpuppies, and waterdogs can usually be kept under aquarium conditions that would be suitable for freshwater fishes. Most terrestrial species will need a moist terrarium with mosses, flat rocks, and bark for hiding. Heating is not the problem with sala-

Children can be taught to appreciate salamanders, especially when other children are the teachers. Tiger salamanders (*left*) are attractive and harmless.

Searching for salamanders in the wild can be a fun and educational family activity.

manders that it can be with pet reptiles. Cool, humid conditions are generally suitable for most salamanders, and sunlight or ultraviolet lamps are not necessary. Most salamander species have not been bred successfully in captivity, in part because so many of them spend most of their lives in one habitat and breed in another.

All salamanders are carnivorous, both as larvae and as adults. Depending on their life stage, they eat primarily aquatic or terrestrial invertebrates. Most will eat crickets, mealworms, and earthworms that can be found at bait stores and pet shops. Dietary supplements may be required for long-term care, although specific information is lacking.

A CONSERVATION VISION FOR EASTERN SALAMANDERS

Salamanders belong to an ancient lineage more than 200 million years old. The ancestors of modern salamanders survived the dinosaur extinctions. In other words, the salamanders we see today are the descendents of animals that successfully reproduced and left offspring through many millennia, surviving changing climates and escaping predators of all types. Copies of many of the same genes that occurred in those salamanders from eons ago still exist today in the cells of the salamander you find under a log or see swimming in a pond. Unfortunately, scientific studies begun several decades ago are showing that many of the prehistoric lineages now represented by modern species will not survive human impacts to their environments. Several species have already been declared extinct, and many more will follow.

Five massive extinction episodes have occurred in the history of the planet, all as a result of natural changes or disasters. Are we to be the cause of the sixth great extinction? Some scientists think so—and think that it will happen in our lifetimes or those of our children. The process has already begun. What can we do about it? Should we protect these animals on moral or practical grounds? One moral argument is that we do not have the right to cause the extinction of another life form, especially one that does not harm us in any way. Another is that an encounter with a colorful salamander is a soul-satisfying experience that connects us to the spirit in nature. Practically speaking, chemical derivatives from the secretions of salamander glands might someday create a wonder drug that cures certain forms of cancer or other diseases. Salamanders already perform valuable ecological services that benefit humans, such as consuming larval mosquitoes and being very important components of natural food chains.

The landscape of the eastern United States is changing for all of its inhabitants. The old farms, plantations, and timberlands are gone, and urban sprawl

above Salamanders that breed in ponds live most of their lives underground in nearby forests and often move up to a thousand yards to find suitable refugia.

right Many salamanders are killed by "urban wildlife" such as raccoons.

has eaten up countless square miles. Human populations are growing in many states in the region, particularly in the Southeast; with them come road networks, impoundments for drinking water, sewer systems, stream channelizations, and more extensive recreational use of natural areas. Loss of habitat is the primary and most widespread effect of human activities—and it is directly responsible for the loss of salamander populations. Some 17 million acres have been converted to urban and suburban environments in the United States since the 1950s. The human population of Florida increased from 2.7 million in 1950 to 22 million in 2022, and most of those people live in expanding population centers. The World Wildlife Fund estimates that wildlife populations in the United States have declined by 68% since 1970.

left Urban sprawl has destroyed old farms, plantations, wetlands, and timberlands throughout the eastern United States.

below Salamanders living in isolated patches of habitat cannot migrate to other fragments across the matrix of barriers and inhospitable lands that humans have created.

Intensive urbanization and untold miles of roads not only destroy natural habitats, they also fragment them into smaller and smaller patches. Salamanders living in these patches are unable to migrate to other fragments of their original population through the matrix of barriers and inhospitable land. Many of those that try are killed while crossing roads. The resulting smaller and smaller populations are more vulnerable to inbreeding, catastrophic weather events, and disease. Small populations in small patches are killed by feral cats and raccoons that are subsidized by human resources. Thus, urbanization and road construction compound a host of other factors that collectively stack the deck against any salamander population living in a habitat fragment. Similar scenarios also occur in agricultural and timberlands. The striped newt, for example, has declined in places where it was once abundant as a result of loss of forest cover, drainage of wetlands, introduction of fish, and fire suppression. Only 15 populations were known to exist in the first decade of the twenty-first century.

Other important causes of salamander population declines include pollution from agriculture, industry, and private lawns; emergent infectious diseases that were not known a few decades ago; and invasive species. Chemicals in agricultural fertilizers, such as atrazine, have been shown to cause mortality and deformities in salamanders and frogs. Chytrid fungus, ranaviruses, and other disease-causing organisms unknown to science until near the end of the twentieth century are now implicated as the cause of amphibian population declines in many areas. Populations of spotted salamanders in western North Carolina have been severely reduced by disease. Several species of earthworms brought to this region have displaced native earthworms and altered the soil humus–leaf litter characteristics of forests, rendering this microhabitat inhospitable for many of the woodland salamanders in the genus *Plethodon*. Invasive plants such as cogongrass also alter soil structure.

Global climate change is predicted to cause warmer temperatures, more wildfires, and changes in rainfall—including longer periods of drought—across the eastern United States. Warmer temperatures are expected to drive mountain salamanders to higher and higher elevations. If that happens, the highest-elevation species, such as the northern pygmy salamander and pygmy salamander, will have no place to go. Longer periods of drought will reduce water levels in ponds and may eliminate the seasonal wetlands and other ephemeral pools mole salamanders require for breeding.

We have it within our power to reduce these human-caused threats and find ways to ameliorate their effects on salamanders. Initial efforts to protect amphibians from extinction started with the passage of federal and state regulations, the most effective being the U.S. Endangered Species Act. Such laws safeguard against commercial collection and killing but, except for the federal act, provide only marginal protection against habitat loss. In addition to the U.S. Fish and Wildlife Service, which administers the endangered and threatened species list, the wildlife agencies within each state have their own lists of species that warrant protection. Efforts to protect sal-

> **DID YOU KNOW?**
>
> Diseases are emerging in increasing numbers of salamander species and populations in the Southeast and worldwide. The chytrid fungus, for example, has caused declines of salamanders in the American West and extinctions of species in the tropics.

A female spotted salamander killed during the breeding season by a raccoon

Habitat restoration and creation are increasingly being used to preserve local populations of salamanders. This created wetland is in eastern Tennessee.

amander species on federal and state lists, and adding other species to such lists, will likely be a part of the salamander conservation tool box for some time to come.

We cannot rely on government regulations to protect all eastern salamanders. Each of us can and should play a role in habitat conservation and protection. The focus should be on habitats where these animals live, because without viable habitat all other protection is futile. Habitat management guidelines exist for all eastern habitats (see Further Reading) and are readily available to everyone. Landowners can manage the habitats on their property to protect species already there or even attract local species by finding ways to connect isolated fragments of habitat. Protecting natural wetlands of all types from pollution, ditching, and filling benefits both salamanders and frogs; however, it is also essential to protect and manage the surrounding upland—usually forested—habitat. Salamanders may breed in ponds, but many live most of their lives underground in the forest and often move up to a thousand yards away to find suitable refugia. Thus, the combined wetland and upland habitat unit should be the target of conservation efforts. Learn what types of habitats are used by salamanders on your land. Work with knowledgeable persons to find ways to make your property more salamander friendly.

Habitat restoration and creation are increasingly being used to ensure that local populations of species do not go extinct. Effective ways to restore natural wetlands are in place (see Further Reading). The Natural Resources Conservation Service (NRCS) branch in every county is staffed with people well versed in wetland resto-

ration. In some cases, it may be advisable to relocate a number of salamanders from a large population to another area to reestablish a population that has been lost. Opportunities to modify degraded lands in ways that result in establishment and restoration of natural habitats are everywhere. Many of these efforts are still trial and error, but we must start somewhere to restore places that once supported salamander populations. Information may be found on websites or in your local college or NRCS office. The most far-reaching and effective organization focusing on conserving all U.S. salamanders and their habitats is Partners in Amphibian and Reptile Conservation (PARC), established in 1999.

The eastern United States, especially the Southeast, includes numerous hotspots of salamander diversity. There are ample opportunities to learn about the 120—well, 119 since the Bay Springs salamander is now extinct—salamanders living in this region. These diverse and beautiful amphibians cannot be saved without your help. Your contributions to efforts to conserve and manage habitats, to support educational programs, and to educate your local land use planning boards—and indeed all of your elected officials—are a good way to be involved as concerned citizens. If you own land, show your neighbors the value of protecting and managing natural habitats. Conservation of these ancient amphibian lineages requires interactions among partners from all walks of life. And that means you.

What Kinds of Salamanders Are Found in Your State?

COMMON AND SCIENTIFIC NAMES	AL	CT	DE	FL	GA	IL	IN
MOLE SALAMANDERS							
Streamside Salamander *Ambystoma barbouri*							■
Reticulated Flatwoods Salamander *Ambystoma bishopi*	■			■	■		
Frosted Flatwoods Salamander *Ambystoma cingulatum*				■	■		
Jefferson Salamander *Ambystoma jeffersonianum*			■			■	■
Blue-spotted Salamander *Ambystoma laterale*			■			■	■
Mabee's Salamander *Ambystoma mabeei*							
Spotted Salamander *Ambystoma maculatum*	■	■	■	■	■	■	■
Marbled Salamander *Ambystoma opacum*	■	■	■	■	■	■	■
Mole Salamander *Ambystoma talpoideum*	■			■	■		
Small-mouthed Salamander *Ambystoma texanum*	■					■	■
Tiger Salamander *Ambystoma tigrinum*	■		■	■	■	■	■
Ambystoma unisex complex			■			■	■
AMPHIUMAS							
Two-toed Amphiuma *Amphiuma means*	■			■	■		
One-toed Amphiuma *Amphiuma pholeter*	■			■	■		
Three-toed Amphiuma *Amphiuma tridactylium*	■						
GIANT SALAMANDERS							
Hellbender *Cryptobranchus alleganiensis*	■				■	■	■
LUNGLESS SALAMANDERS							
Green Salamander *Aneides aeneus*	■				■		■
Hickory Nut Gorge Green Salamander *Aneides caryansis*							
Cumberland Dusky Salamander *Desmognathus abditus*							
Cherokee Mountain Dusky Salamander *Desmognathus adatsihi*							
Seepage Salamander *Desmognathus aeneus*	■				■		
Nantahala Black-bellied Salamander *Desmognathus amphileucus*					■		
Foothills Dusky Salamander *Desmognathus anicetus*							
Apalachicola Dusky Salamander *Desmognathus apalachicolae*	■			■	■		
Golden Shovel-nosed Salamander *Desmognathus aureatus*					■		
Southern Dusky Salamander *Desmognathus auriculatus*	■			■	■		
Piedmont Dusky Salamander *Desmognathus bairdi*							
Balsam Mountain Dusky Salamander *Desmognathus balsameus*							
Camp's Dusky Salamander *Desmognathus campi*					■		
Carolina Mountain Dusky Salamander *Desmognathus carolinensis*							
Western Dusky Salamander *Desmognathus catahoula*							
Talladega Seal Salamander *Desmognathus cheaha*	■				■		
Spotted Dusky Salamander *Desmognathus conanti*	■			■	■	■	
Dwarf Blackbelly Salamander *Desmognathus folkertsi*					■		
Northern Dusky Salamander *Desmognathus fuscus*		■	■				■
Cherokee Black-bellied Salamander *Desmognathus gunigeusgwotli*							
Imitator Salamander *Desmognathus imitator*							
Western Shovel-nosed Salamander *Desmognathus intermedius*							
Kanawha Black-bellied Salamander *Desmognathus kanawha*							
Wolf Dusky Salamander *Desmognathus lycos*							
Shovel-nosed Salamander *Desmognathus marmorata*							
Pisgah Black-bellied Salamander *Desmognathus mavrokoilius*							
Seal Salamander *Desmognathus monticola*	■				■		
Allegheny Mountain Salamander *Desmognathus ochrophaeus*							
Ocoee Salamander *Desmognathus ocoee*	■				■		
Blue Ridge Dusky Salamander *Desmognathus orestes*							

	KY	LA	ME	MD	MA	MI	MN	MS	NH	NJ	NY	NC	OH	PA	RI	SC	TN	VT	VA	WV	WI	TOTAL
AMBYSTOMIDAE																						
	■												■				■			■		5
																						3
														■								3
	■			■	■				■	■	■		■	■			■	■	■			14
			■		■	■	■		■	■	■		■	■				■			■	14
												■	■					■				3
	■	■	■	■	■	■	■	■	■	■	■	■	■	■	■	■	■	■	■	■	■	28
	■	■		■	■	■		■	■	■	■	■	■	■	■	■	■	■	■			25
	■	■						■				■	■				■	■				11
	■	■					■			■							■			■		10
	■	■		■		■	■	■		■	■	■	■			■	■	■			■	20
	■				■		■	■		■	■		■	■				■			■	14
AMPHIUMIDAE																						
		■		■								■				■			■			8
				■																		4
	■	■		■													■					5
CRYPTOBRANCHIDAE																						
	■			■				■			■	■	■	■			■		■	■		14
PLETHODONTIDAE																						
	■			■				■				■	■	■		■	■		■	■		13
												■										1
																	■					1
																	■					1
												■				■	■					5
												■				■	■					4
												■				■						2
												■					■					3
												■					■					3
												■					■					3
												■					■	■				3
												■										1
												■					■					3
												■					■					2
		■																				1
																						2
	■	■						■				■					■					9
												■				■						3
	■			■	■	■				■	■	■	■	■	■		■	■	■	■		18
												■					■					2
												■					■					2
												■					■					2
												■							■	■		3
												■			■		■		■			4
												■					■		■	■		4
												■					■		■			3
	■			■								■		■		■	■		■	■		10
	■			■						■	■		■	■			■		■	■		9
												■					■					4
	■																■		■			3

COMMON AND SCIENTIFIC NAMES	AL	CT	DE	FL	GA	IL	IN
LUNGLESS SALAMANDERS							
Northern Pygmy Salamander *Desmognathus organi*							
Pascagoula Dusky Salamander *Desmognathus pascagoula*	■						
Chattooga Dusky Salamander *Desmognathus perlapsus*	■				■		
Flat-headed Dusky Salamander *Desmognathus planiceps*							
Black-bellied Salamander *Desmognathus quadramaculatus**							
Santeetlah Dusky Salamander *Desmognathus santeetlah*							
Tilley's Dusky Salamander *Desmognathus tilleyi*							
Valentine's Southern Dusky Salamander *Desmognathus valentinei*	■						
Carolina Swamp Dusky Salamander *Desmognathus valtos*					■		
Black Mountain Salamander *Desmognathus welteri*							
Pygmy Salamander *Desmognathus wrighti*					■		
Brownback Salamander *Eurycea aquatica*	■				■		
Carolina Sandhills Salamander *Eurycea arenicola*							
Northern Two-lined Salamander *Eurycea bislineata*		■	■				
Chamberlain's Dwarf Salamander *Eurycea chamberlaini*							
Southern Two-lined Salamander *Eurycea cirrigera*	■			■	■	■	■
Three-lined Salamander *Eurycea guttolineata*	■			■	■		
Hillis's Dwarf Salamander *Eurycea hillisi*	■			■	■		
Junaluska Salamander *Eurycea junaluska*							
Long-tailed Salamander *Eurycea longicauda*	■		■		■	■	■
Cave Salamander *Eurycea lucifuga*	■				■	■	■
Western Dwarf Salamander *Eurycea paludicola*							
Dwarf Salamander *Eurycea quadridigitata*	■			■	■		
Bog Dwarf Salamander *Eurycea sphagnicola*	■			■			
Georgia Blind Salamander *Eurycea wallacei*				■	■		
Blue Ridge Two-lined Salamander *Eurycea wilderae*					■		
Berry Cave Salamander *Gyrinophilus gulolineatus*							
Spring Salamander *Gyrinophilus porphyriticus*	■	■			■		
Tennessee Cave Salamander *Gyrinophilus palleucus*	■				■		
West Virginia Spring Salamander *Gyrinophilus subterraneus*							
Four-toed Salamander *Hemidactylium scutatum*	■	■	■	■	■	■	■
Red Hills Salamander *Phaeognathus hubrichti*	■						
Blue Ridge Gray-cheeked Salamander *Plethodon amplus*							
Tellico Salamander *Plethodon aureolus*							
Chattahochee Slimy Salamander *Plethodon chattahoochee*					■		
Cheoah Bald Salamander *Plethodon cheoah*							
Atlantic Coast Slimy Salamander *Plethodon chlorobryonis*					■		
Eastern Red-backed Salamander *Plethodon cinereus*		■	■			■	■
White-spotted Slimy Salamander *Plethodon cylindraceus*							
Dixie Caverns Salamander *Plethodon dixi*							
Northern Zigzag Salamander *Plethodon dorsalis*	■					■	■
Northern Ravine Salamander *Plethodon electromorphus*							■
Northern Slimy Salamander *Plethodon glutinosus*	■	■	■		■	■	■
Southeastern Slimy Salamander *Plethodon grobmani*	■			■	■		
Valley and Ridge Salamander *Plethodon hoffmani*							
Peaks of Otter Salamander *Plethodon hubrichti*							
Blacksburg Salamander *Plethodon jacksoni*							
Red-cheeked Salamander *Plethodon jordani*							

* The black-bellied salamander (*Desmognathus quadramaculatus*), formerly recognized as a single species, has been determined to be a complex of four genetically distinct salamander species— *D. gvnigeusgwotli, D. mavrokoileus, D. kanawha,* and *D. amphileucus.*

KY	LA	ME	MD	MA	MI	MN	MS	NH	NJ	NY	NC	OH	PA	RI	SC	TN	VT	VA	WV	WI	TOTAL
PLETHODONTIDAE *(continued)*																					
											■					■		■			3
							■														2
											■				■						4
																		■			1
																					0
											■				■						2
											■				■						2
	■						■														3
											■				■						3
■															■			■	■		4
											■				■						3
							■				■				■						4
											■										1
		■		■	■			■	■	■	■		■	■		■	■	■			14
											■				■						2
■	■						■				■		■		■	■		■	■		14
■	■						■				■				■	■		■			10
											■				■						3
											■				■						2
■				■			■			■	■	■	■		■			■	■		16
■							■				■		■		■			■	■		10
	■						■														2
											■				■						5
							■														3
																					2
											■				■	■		■			5
															■						1
■			■	■	■		■	■	■	■	■	■	■	■	■	■	■	■	■		20
															■						3
																			■		1
■	■	■	■	■	■	■	■	■	■	■	■	■	■	■	■	■	■	■	■	■	28
											■										1
											■										1
											■					■					2
											■										2
											■										1
													■					■			4
■		■	■	■	■	■		■	■	■	■	■	■	■		■		■	■	■	22
											■				■	■		■	■		5
																		■			1
■															■						5
■												■	■						■		5
■			■					■	■	■	■	■	■		■			■	■		17
																					3
			■										■					■	■		4
																		■			1
										■								■			2
											■					■					2

	AL	CT	DE	FL	GA	IL	IN
LUNGLESS SALAMANDERS							
Cumberland Plateau Salamander *Plethodon kentucki*							
Louisiana Slimy Salamander *Plethodon kisatchie*							
Crevice Salamander *Plethodon longricus*							
South Mountain Gray-cheeked Salamander *Plethodon meridianus*							
Southern Gray-checked Salamander *Plethodon metcalfi*					■		
Mississippi Slimy Salamander *Plethodon mississippi*	■						
Northern Gray-cheeked Salamander *Plethodon montanus*							
Cheat Mountain Salamander *Plethodon nettingi*							
Ocmulgee Slimy Salamander *Plethodon ocmulgee*					■		
Yellow-spotted Woodland Salamander *Plethodon pauleyi*							
Pigeon Mountain Salamander- *Plethodon petraeus*					■		
Cow Knob Salamander *Plethodon punctatus*							
Southern Ravine Salamander *Plethodon richmondi*							
Savannah Slimy Salamander *Plethodon savannah*					■		
Southern Redback Salamander *Plethodon serratus*	■				■		
Shenandoah Salamander *Plethodon shenandoah*							
Big Levels Salamander *Plethodon sherando*							
Red-legged Salamander *Plethodon shermani*					■		
Southern Appalachian Salamander *Plethodon teyahalee*					■		
South Carolina Slimy Salamander *Plethodon variolatus*					■		
Southern Zigzag Salamander *Plethodon ventralis*	■						
Shenandoah Mountain Salamander *Plethodon virginia*							
Webster's Salamander *Plethodon websteri*	■				■		
Wehrle's Salamander *Plethodon wehrlei*							
Weller's Salamander *Plethodon welleri*							
Yonahlossee Salamander *Plethodon yonahlossee*							
Mud Salamander *Pseudotriton montanus*	■		■	■	■		
Red Salamander *Pseudotriton ruber*	■		■	■	■		■
Many-lined Salamander *Stereochilus marginatus*				■	■		
Patch-nosed Salamander *Urspelerpes brucei*					■		
WATERDOGS AND MUDPUPPIES							
Black Warrior River Waterdog *Necturus alabamensis*	■						
Gulf Coast Waterdog *Necturus beyeri*	■			■	■		
Neuse River Waterdog *Necturus lewisi*							
Common Mudpuppy *Necturus maculosus*	■	■			■	■	■
Apalachicola Waterdog *Necturus moleri*	■			■	■		
Escambia Waterdog *Necturus mounti*	■			■	■		
Dwarf Waterdog *Necturus punctatus*					■		
NEWTS							
Striped Newt *Notophthalmus perstriatus*				■	■		
Eastern Newt *Notophthalmus viridescens*	■	■	■	■	■	■	■
SIRENS							
Southern Dwarf Siren *Pseudobranchus axanthus*				■			
Northern Dwarf Siren *Pseudobranchus striatus*				■	■		
Lesser Siren *Siren intermedia*	■			■	■		
Greater Siren *Siren lacertina*	■			■	■		
Western Lesser Siren *Siren nettingi*	■					■	■
Reticulated Siren *Siren reticulata*	■			■			
Seepage Siren *Siren spagnicola*	■			■			
STATE SPECIES TOTAL	52	13	12	33	64	20	22

KY	LA	ME	MD	MA	MI	MN	MS	NH	NJ	NY	NC	OH	PA	RI	SC	TN	VT	VA	WV	WI	TOTAL
PLETHODONTIDAE (continued)																					
▪																▪		▪	▪		4
	▪																				1
											▪										1
											▪										1
											▪				▪						3
▪	▪						▪									▪					5
											▪					▪			▪		3
																			▪		1
																					1
▪																▪		▪			4
																					1
																		▪	▪		2
▪											▪					▪		▪	▪		5
																					1
											▪					▪		▪			5
																			▪		1
																			▪		1
															▪	▪					3
															▪						4
															▪						2
▪							▪				▪							▪	▪		7
																		▪	▪		2
							▪								▪			▪			5
			▪							▪		▪						▪	▪		5
											▪					▪		▪			3
											▪					▪		▪			3
▪	▪			▪			▪		▪		▪	▪	▪		▪	▪		▪	▪		16
▪	▪			▪			▪		▪	▪	▪	▪	▪		▪	▪		▪	▪		18
											▪					▪		▪			5
															▪						2
PROTEIDAE																					
																					1
	▪						▪														5
											▪										1
▪	▪	▪	▪	▪	▪	▪	▪				▪	▪	▪	▪		▪	▪	▪	▪	▪	24
																					3
																					2
															▪			▪			4
SALAMANDRIDAE																					
																					2
▪	▪	▪	▪	▪	▪	▪	▪	▪	▪	▪	▪	▪	▪	▪	▪	▪	▪	▪	▪	▪	28
SIRENIDAE																					
																					1
															▪						3
							▪								▪			▪			7
							▪								▪			▪			6
▪	▪				▪		▪								▪						8
																					2
	▪						▪														4
39	**25**	**10**	**20**	**12**	**12**	**7**	**32**	**12**	**17**	**19**	**73**	**26**	**23**	**9**	**40**	**65**	**12**	**57**	**37**	**8**	

What Salamanders in Your State Are Protected?

COMMON AND SCIENTIFIC NAMES	AL	CT	DE	FL	GA	IL	IN
MOLE SALAMANDERS							
Streamside Salamander *Ambystoma barbouri*						SC	SC
Reticulated Flatwoods Salamander *Ambystoma bishopi*	FE			FE	FE		
Frosted Flatwoods Salamander *Ambystoma cingulatum*				FT	FT		
Jefferson Salamander *Ambystoma jeffersonianum*		SC					
Blue-spotted Salamander *Ambystoma laterale*		SE				SC	SC
Mabee's Salamander *Ambystoma mabeei*							
Spotted Salamander *Ambystoma maculatum*							
Marbled Salamander *Ambystoma opacum*							
Mole Salamander *Ambystoma talpoideum*						SE	SE
Small-mouthed Salamander *Ambystoma texanum*							
Tiger Salamander *Ambystoma tigrinum*				SE			
Ambystoma unisex complex		SC			SE		
AMPHIUMAS							
Two-toed Amphiuma *Amphiuma means*	SC						
One-toed Amphiuma *Amphiuma pholeter*	CI				SR		
Three-toed Amphiuma *Amphiuma tridactylium*	SC						
GIANT SALAMANDERS							
Hellbender *Cryptobranchus alleganiensis*	CI				ST	SE	SE
LUNGLESS SALAMANDERS							
Green Salamander *Aneides aeneus*	SC				SR		SE
Hickory Nut Gorge Green Salamander *Aneides caryansis*							
Cumberland Dusky Salamander *Desmognathus abditus*							
Seepage Salamander *Desmognathus aeneus*	SC						
Apalachicola Dusky Salamander *Desmognathus apalachicolae*	SC						
Southern Dusky Salamander *Desmognathus auriculatus*	SC						
Talladega Seal Salamander *Desmognathus cheaha*	SC						
Dwarf Black-bellied Salamander *Desmognathus folkertsi*							
Seal Salamander *Desmognathus monticola*	SC						
Ocoee Salamander *Desmognathus ocoee*	SC						
Black Mountain Salamander *Desmognathus welteri*							
Pygmy Salamander *Desmognathus wrighti*							
Brown-backed Salamander *Eurycea aquatica*					SC		
Three-lined Salamander *Eurycea guttolineata*							
Junaluska Salamander *Eurycea junaluska*							
Long-tailed Salamander *Eurycea longicauda*							
Cave Salamander *Eurycea lucifuga*							
Dwarf Salamander *Eurycea quadridigitata*							
Georgia Blind Salamander *Eurycea wallacei*				ST	ST		
Berry Cave Salamander *Gyrinophilus gulolineatus*							
Spring Salamander *Gyriniphilus porphyriticus*		ST					

CI State listed as Critically Imperiled

SC State listed as Special Concern (includes "In Need of Management" and "High Priority")

SR State listed as Rare

SP State listed as Protected from Taking or Possessing

KY	LA	ME	MD	MA	MI	MN	MS	NH	NJ	NY	NC	OH	PA	RI	SC	TN	VT	VA	WV	WI
AMBYSTOMIDAE																				
																SE			SP	
															FT					
					SC		SC									SC			SP	
					ST	SE	SC				SE	SE				SC				
											ST							ST		
			SC																SP	
					ST	SE		SE	SC	SC									SP	
											SC									
						SE													SP	
	SP		SE					SE	SE	SE	ST				SC			SE		
					SC		SC													
AMPHIUMIDAE																				
							SE													
SE																				
CRYPTOBRANCHIDAE																				
SC			SE	SE					SC	SC	SE	SC				ST		SC	SP	
PLETHODONTIDAE																				
SC			SE	SE							ST	SE	ST		SC				SP	
											SE									
SC																SC				
															SC	SC				
															SC					
SC																			SP	
															SC					
ST																				
											ST					SC				
									ST	SC	ST								SP	
				SE								SE							SP	
											SC									
																ST				
		SC		SE				SC							SC				SP	

COMMON AND SCIENTIFIC NAMES	AL	CT	DE	FL	GA	IL	IN
LUNGLESS SALAMANDERS							
Tennessee Cave Salamander *Gyriniphilus palleucus*	SC				ST		
West Virginia Spring Salamander *Gyrinophilus subterraneus*							
Four-toed Salamander *Hemidactylium scutatum*						SC	SC
Red Hills Salamander *Phaeognathus hubrichti*	FT						
Eastern Red-backed Salamander *Plethodon cinereus*							
Northern Ravine Salamander *Plethodon electromorphus*							
Northern Slimy Salamander *Plethodon glutinosus*		ST					
Valley and Ridge Salamander *Plethodon hoffmani*							
Peaks of Otter *Plethodon hubrichti*							
Blacksburg Salamander *Plethodon jacksoni*							
Cumberland Plateau Salamander *Plethodon kentucki*							
Louisiana Slimy Salamander *Plethodon kisatchie*							
Cheat Mountain Salamander *Plethodon nettingi*							
Yellow-spotted Woodland Salamander *Plethodon pauleyi*							
Pigeon Mountain Salamander *Plethodon petraeus*					SR		
Cow Knob Salamander *Plethodon punctatus*							
Southern Ravine Salamander *Plethodon richmondi*							
Southern Redback Salamander *Plethodon serratus*	SC						
Shenandoah Salamander *Plethodon shenandoah*							
Big Levels Salamander *Plethodon sherando*							
Red-legged Salamander *Plethodon shermani*					SC		
Southern Appalachian Salamander *Plethodon teyahalee*							
Shenandoah Mountain Salamander *Plethodon virginia*							
Webster's Salamander *Plethodon websteri*							
Wehrle's Salamander *Plethodon wehrlei*							
Weller's Salamander *Plethodon welleri*							
Mud Salamander *Pseudotriton montanus*			SE				
Red Salamander *Pseudotriton ruber*						SE	SE
Patch-nosed Salamander *Urspelerpes brucei*					SC		
WATERDOGS AND MUDPUPPIES							
Black Warrior River Waterdog *Necturus alabamensis*	FE						
Gulf Coast Waterdog *Necturus beyeri*							
Neuse River Waterdog *Necturus lewisi*							
Common Mudpuppy *Necturus maculosus*	CI	SC			SC	SE	SC
NEWTS							
Striped Newt *Notophthalmus perstriatus*				ST	ST		
Eastern Newt *Notophthalmus viridescens*							
SIRENS							
Northern Dwarf Siren *Pseudobranchus striatus*							
Greater Siren *Siren lacertina*	SC						
Western Lesser Siren *Siren nettingi*							

KY	LA	ME	MD	MA	MI	MN	MS	NH	NJ	NY	NC	OH	PA	RI	SC	TN	VT	VA	WV	WI
PLETHODONTIDAE (continued)																				
																ST				
																			SP	
SC							SC				SC	SC			SC	SC	SC		SP	SC
SC																			SP	
																			SP	
																			SP	
																			SP	
																		SC		
											ST									
																			SP	
	CI																			
																			FT	
SC																SC			SP	
																SC			SP	
																			SP	
																		FE		
																			SP	
											SC									
																			SP	
	CI														SE					
					SC														SP	
											SC					SC		SC		
									ST			ST	SE		SC				SP	
																			SP	
															SC					
PROTEIDAE																				
	SR																			
											FT									
				SE			SC				ST							SC	SP	
SALAMANDRIDAE																				
																			SP	
SIRENIDAE																				
																ST				
SC						SC														

GLOSSARY

Adelgid An exotic insect (Hemiptera: Adelgidae: *Adelges tsugae*) introduced from Asia that has affected salamander populations in the Appalachian Mountains by killing hemlocks and Fraser firs.

Aestivation A period of inactivity during dry and/or hot periods in which animals wait for conditions suitable for feeding and other activities to return.

Albino An animal completely lacking the pigment that provides the color to skin and eyes. Animals lacking only dark pigment, or melanin, are often referred to as albinos but are more properly said to be amelanistic.

Anterior Referring to the end of the animal toward the head.

Anurans Amphibians that are frogs or toads.

Autotomy (*v.* autotomize) The breaking off of an appendage, such as the tail, as a means of escaping predation.

Biodiversity The number, distribution, and abundance of species within a given area.

Bioindicator A species whose health or condition, either at the individual level or at the population level, indicates the condition of the habitat or ecosystem as a whole.

Biomass The weight of living things in the environment; or the weight of all the animals in a population.

Biomonitor To measure and record the number, types, and characteristics of organisms.

Canthus rostralis A dark-bordered white line that runs from the eye to the nostril of the spring salamander (*Gyrinophilus porphyriticus*).

Caudates Amphibians that are salamanders.

Chytrid A fungus that many herpetologists consider responsible for the recent decline—and in some instances the extinction—of frog and toad populations, especially in high-elevation tropical habitats.

Chytridiomycosis The disease caused by the chytrid fungus that is sometimes fatal to amphibians.

Cirri (*sing.* cirrus) A pair of fleshy projections that point downward from the upper lip of some salamanders in the genus *Eurycea*. Cirri are most prevalent in males during the breeding season but are present in females of some species.

Cloaca (*adj.* cloacal) A single opening through which the urinary, digestive, and reproductive tracts exit the body.

Clutch A group of eggs laid together at one time by a single individual.

Cold-blooded A nontechnical term that refers to animals whose body temperature is largely determined by environmental conditions and the thermoregulatory behavior of the animal (*see* Ectotherm).

Courtship The series of behaviors that precedes mating.

Cryptic species Species that are genetically different but cannot be readily distinguished morphologically.

Desmogs A colloquial term for dusky salamanders belonging to the genus *Desmognathus*.

Dorsolateral Referring to the area of the body between the back and sides.

Ecology The study of how organisms interact with their environment and other organisms.

Ectotherm (*adj.* ectothermic) An animal whose body temperature is largely determined by environmental conditions and the thermoregulatory behavior of the animal (*see* Cold-blooded).

Elastomer Material injected under the skin of amphibians as a colored liquid that hardens and becomes a visible marker that can be used to identify individuals.

Electromorphs Species that have been described as distinct from closely related species that are similar in appearance on the basis of genetic differences.

Endangered Referring to a species or population that is considered at risk of becoming extinct.

Endotherm An animal that maintains a constant body temperature primarily through the use of heat generated by its own metabolism (*see* Warm-blooded).

Ephemeral wetland Seasonal wetland habitats that usually hold water for only certain periods of the year.

Extinct Referring to species with no living individuals.

Fall Line The line separating the Piedmont from the Coastal Plain in the eastern United States.

Family A taxonomic group containing two or more closely related genera.

Generalist An animal that does not specialize on any particular type of prey or is not restricted to a particular habitat.

Genus (pl. genera) A taxonomic grouping of one or more closely related species.

Herpetofauna The amphibians and reptiles that inhabit a given area.

Herpetologist A scientist who studies amphibians and reptiles.

Herpetology The scientific study of amphibians and reptiles.

Hibernaculum The microhabitat to which a salamander retreats during cold weather.

Hibernation A period of inactivity during cold periods; also known as "brumation" when referring to amphibians and reptiles.

Hybrid The offspring of mating between two different species.

Internal nares Openings in the roof of the mouth of salamanders that lead to the nasal passage.

Intergrade An intermediate form of a species resulting from mating and genetic mixing between individuals of two or more subspecies within a zone where their ranges overlap. Intergrade individuals may possess traits of two or more subspecies.

Mental gland A disk- to U-shaped gland on the underside of the chin of some plethodontid salamander males that produces pheromones used during courtship to entice females to mate.

Metamorphosis The process by which a larval aquatic salamander changes into a terrestrial juvenile.

Metamorph An amphibian that has recently undergone metamorphosis.

Microhabitat A small, defined area within the general habitat.

Nymph An aquatic insect larva.

Neoteny (adj. neotenic) A condition in which an adult salamander has certain traits such as gills that are typically found in aquatic larvae; *also* paedomorphosis.

Nocturnal Active at night.

Orthopteran A member of the insect order Orthoptera, which includes the katydids, grasshoppers, and crickets.

Oxbow A U-shaped lake that was once the main channel of a nearby river.

Paedomorphosis *See* Neoteny.

PARC Acronym for Partners in Amphibian and Reptile Conservation, the largest partnership group dedicated to the conservation of all amphibians and reptiles and their habitats.

Phylogeny The evolutionary and genealogical relationships among different groups and species of animals.

Physiographic region An area defined by the geological structure and general topography of the landscape.

PIT tag Acronym for "passive integrated transponder tag," a glass-encapsulated electronic device injected into the body cavity of animals for identification purposes.

Plethodontid (*adj.*) Referring to the family Plethodontidae.

Posterior Away from the head of an animal and toward the tail.

Radiotelemetry A method using a radiotransmitter attached to or implanted in an animal to track its movements by locating it using a directional antenna and radio receiver.

Reticulate A network pattern on the skin formed by an interconnecting series of light lines.

Rugose Creased or wrinkled.

Seep A permanently wet area created by groundwater slowly flowing out at the base of a steep bluff or between bedrock formations.

Silviculture Planned and managed growth and development of forests.

Specialist An animal restricted in its choice of diet or habitat.

Species Typically an identifiable and distinct group of organisms whose members breed and produce viable offspring under natural conditions.

Spermatophore A capsule of spermatozoa atop a gelatinous stalk deposited during courtship by the male of some species of salamanders.

Spring lizard A colloquial term for various species of southeastern mountain salamanders, particularly the dusky salamanders, that are used for fish bait.

Subspecies A taxonomic unit, or "race," within a species, usually defined as morphologically distinct and occupying a geographic range that does not overlap with that of other races of the species. Subspecies may interbreed naturally in areas of geographic contact (*see* Intergrade).

Sympatric Referring to different species that live in the same general geographic range.

Taxonomy The scientific field of classification and naming of organisms.

Troglobite A cave-dwelling animal.

Vaccination The process in which the male of some plethodontid salamander species uses his enlarged front teeth to scratch the skin of a female and then presses his mental gland against the abraded site to release a secretion that enters the female's body and presumably makes her receptive for mating.

Warm-blooded A nontechnical term referring to an animal that maintains its body temperature primarily through the use of metabolic heat (*see* Endotherm).

Woodland salamander A general term for all of the species in the genus *Plethodon*, which are terrestrial and have no aquatic larval stage.

FURTHER READING

Altig, R. and R. McDiarmid. 2015. *Handbook of Larval Amphibians of the United States and Canada*. Ithaca, N.Y.: Comstock Publishing Associates.

Bailey, M. A., J. N. Holmes, K. A. Buhlmann, and J. C. Mitchell. 2006. *Habitat Management Guidelines for Amphibians and Reptiles of the Southeastern United States*. Montgomery, Ala.: Partners in Amphibian and Reptile Conservation. Tech. Publ. HMG-2.

Barbour, R. W. 1971. *Amphibians and Reptiles of Kentucky*. Lexington: University Press of Kentucky.

Bartlett, P. P., and R. D. Bartlett. 2006. *Guide and Reference to the Amphibians of Eastern and Central North America (North of Mexico)*. Gainesville: University of Florida Press.

Bartlett, R. D., and P. P. Bartlett. 1999. *A Field Guide to Florida Reptiles and Amphibians*. Houston: Gulf Publishing Company.

Bean, J., A. Braswell, J. C. Mitchell, J. Harrison, and W. Palmer. 2010. *Amphibians and Reptiles of the Carolinas and Virginia*. 2nd ed. Chapel Hill: University of North Carolina Press.

Bounder, J., and J. L. Carr. 2017. *Amphibians and Reptiles of Louisiana: An Identification and Reference Guide*. Baton Rouge: Louisiana State University Press.

Carmichael, P., and W. Williams. 2001. *Florida's Fabulous Reptiles and Amphibians*. Tampa, Fla.: World Publications.

Dodd, C. K. Jr. 2004. *The Amphibians of the Great Smoky Mountains National Park*. Knoxville: University of Tennessee Press.

Duellman, W. E., and L. Trueb. 1994. *Biology of Amphibians*. Baltimore: Johns Hopkins University Press.

Dundee, H. A., and D. A. Rossman. 1996. *Amphibians and Reptiles of Louisiana*. Baton Rouge: Louisiana State University Press.

Gibbons, J. W., and R. D. Semlitsch. 1991. *Guide to Reptiles and Amphibians of the Savannah River Site*. Athens: University of Georgia Press.

Jensen, J. B., C. D. Camp, J. W. Gibbons, and M. J. Elliott, eds. 2008. *Amphibians and Reptiles of Georgia*. Athens: University of Georgia Press.

Krysko, K., K. Enge, and P. Moler. 2019. *Amphibians and Reptiles of Florida.* Gainesville: University Press of Florida.

Lannoo, M., ed. 2005. *Amphibian Declines: The Conservation Status of United States Species.* Berkeley: University of California Press.

Mitchell, J. C., A. R. Breisch, and K. A. Buhlmann. 2006. *Habitat Management Guidelines for Amphibians and Reptiles of the Northeastern United States.* Montgomery, Ala.: Partners in Amphibian and Reptile Conservation. Tech. Publ. HMG-2.

Mount, R. M. 1975. *The Reptiles and Amphibians of Alabama.* Tuscaloosa: University of Alabama Press.

Niemiller, M., and R. Reynolds. 2011. *The Amphibians and Reptiles of Tennessee.* Knoxville: The University of Tennessee Press.

Petranka, J. W. 1998. *Salamanders of the United States and Canada.* Washington, D.C.: Smithsonian Institution Press.

Pfingsten, R. A., J. G. Davis, T. O. Matson, G. Lipps Jr., D. Wynn, and B. Armitage, eds. 2013. *Amphibians of Ohio.* Columbus: Ohio Biological Survey.

Powell, R., R. Conant, and J. Collins. 2016. *Peterson Field Guide to Reptiles and Amphibians of Eastern and central North America.* 4th ed. Boston: Houghton Mifflin.

Pyron, R. Alexander, and David A. Beamer. 2022. "Nomenclatural solutions for diagnosing 'cryptic' species using molecular and morphological data facilitate a taxonomic revision of the Black-bellied Salamander (Urodela, *Desmognathus* 'quadramaculatus') from the southern Appalachian Mountains." *Bionomina* 27: 1–43.

———. 2022. "Systematics of the Ocoee Salamander (Plethodontidae: *Desmognathus ocoee*), with descriptions of two new species from the southern Blue Ridge Mountains." *Zootaxa* 51(2): 207–240.

———. 2023. "A systematic revision of the Shovel-nosed Salamander (Plethodontidae: *Desmognathus marmoratus*), with re-description of the related *D. aureatus* and *D. intermedius*." *Zootaxa* 5270 (2): 262–280

———. 2023. "Systematic revision of the Spotted and Northern Dusky Salamanders (Plethodontidae: *Desmognathus conanti* and *Desmognathus fuscus*), with six new species from the eastern United States." *Zootaxa* 5311 (4): 451–504.

Semlitsch, R. D., ed. 2003. *Amphibian Conservation.* Washington, D.C.: Smithsonian Institution Press.

Stebbins, R. C., and N. W. Cohen. 1995. *A Natural History of Amphibians.* Princeton: Princeton University Press.

Wilson, L. A. 1995. *Land Manager's Guide to the Amphibians of the South.* Chapel Hill, N.C.: The Nature Conservancy and Forest Service.

ACKNOWLEDGMENTS

We are grateful to the numerous herpetologists—friends, colleagues, and others—who have engaged in insightful conversations, provided information on ecology, behavior, and localities, and offered useful comments for *Salamanders of the Eastern United States*. This volume expands upon and updates the contents of *Salamanders of the Southeast* (2010), upon which it is based, to include not just the Southeast but the entire eastern United States with all its recently described species. Books like this one cannot be completed without the help of many people.

Initial contributors to the earlier southeastern version included: Mark Bailey, Dick Bartlett, David Beamer, Jeff Beane, Steve Bennett, Jeff Boundy, Alvin Braswell, Dick Bruce, Kurt Buhlmann, Carlos Camp, Erica Crespi, Ken Dodd, Anne R. Gibbons, Gabrielle Graeter, Kristin Grayson, Judy Greene, Susan Lane Harris, Julian Harrison, Jeff Humphries, Bob Jones, Trip Lamb, Tom Luhring, John MacGregor, Bruce Means, Brian Miller, Paul Moler, Matt Niemiller, Tom Pauley, Leslie Rissler, Steve Roble, Travis Ryan, David Scott, Ray Semlitsch, Dirk Stevenson, Brian Todd, Jayme Waldron, Margaret Wead, Patricia West, J. D. Willson, and Addison Wynn. John Jensen read an early draft of the entire manuscript of *Salamanders of the Southeast* and provided excellent suggestions for improvement. Margaret Wead scanned slides and organized digital images for the book and was helpful in a variety of other ways with the organization and preparation of files. Mindy Conner did an excellent job of copyediting the 2010 book and others in the series.

Herpetologists throughout the country provided a spectacular array of salamander slides and images. The excellent color images of species and habitats generously offered by so many experts in animal photography have added tremendously to the value of this book. We thank the following for providing images and slides first used in *Salamanders of the Southeast*: Richard Bartlett, Steve Bennett, E. Pierson Hill, Bill Hopkins, Tom Luhring, David Scott, Brian Todd, R. Wayne VanDevender, and J. D. Willson. Images new to *Salamanders of the Eastern United States* were provided by Jeff Beane, Zach Felix, Dante Fenolio,

Kevin Hutcheson, Jake Hutton, Will Lattea, Bruce Means, Todd Pierson, Jake Scott, Wayne Van Devender, and Larry Wilson.

Whit Gibbons is especially appreciative of the support and understanding of his family during the preparation of the book: Carolyn, Laura, Michael, Jennifer, Allison, Parker, and Gilbey Gibbons; Susan Lane, Keith, and Nicholas Harris; Jennifer, Jim, and Samuel High. Finally, because the book supports the efforts of Partners in Amphibian and Reptile Conservation (PARC) to promote education about reptiles and amphibians, we thank the many PARC members who offered encouragement, advice, and enthusiastic support.

Finally, we thank Melissa B. Buchanan of the University of Georgia Press for her untiring efforts with the editing process of *Salamanders of the Eastern United States*.

Special Acknowledgments by Larry Wilson

I especially thank Paul Moler for escorting me around Florida and its many habitats over almost 45 years and always sharing information, answering many questions, and providing wonderful conversations about all things herpetological, but especially about *Amphiuma*, *Siren*, and *Necturus*. I also thank Wayne Van Devender for providing and sharing his years of wisdom and field knowledge. He is one of the friendliest herpetologists there is. He is always willing to provide photos and data. I really appreciate his kindness and help on many aspects of this book. I also thank Tom Pauley, a good friend for many years and one of the real institutions in the field of salamander biology. Over the years he has mentored over 90 graduate students in the field of herpetology. He was most helpful in escorting me around his home turf of West Virginia and providing localities and information vital to the expansion of this book to the north. J. J. Apodaca was very helpful in escorting me to see *Aneides caryaensis* and *Plethodon longricus* habitats of Hickory Nut Gorge. I appreciate his editing those species accounts and range maps and helping me see these amazing creatures in person.

Other herpetologists have been most helpful in the creation of this book by providing wonderful discussions and generously exchanging information. Finding and exploring the fascinating history and biology of the *Ambystoma* unisexual salamanders was greatly aided by discussions with Fred Kraus, Greg Lipps, and Rob Denton. Thanks to all of them for locality information and to Rob for inviting me into his lab to observe, discuss, and photograph specimens. I was unaccustomed to field work in the "frigid" north, finding unisexual salamanders in snow flurries, high winds, and sleet in April. I thank Dave Beamer for long discussions on cryptic salamander "complexes" and his sharing of insights, localities, and published and unpublished research. My knowledge of cryptic species went way up the learning curve for a field ecologist. Others who were most helpful in keeping us abreast with the new cryptic species (and molecular biology) of taxonomy were Zach Felix, Kevin Hamed, Todd Pierson, and Carlos Camp.

People who provided natural history information and localities that greatly helped out our effort included: Tom Mann (*Plethodon websteri*), Bryan Scott (*Plethodon longricus*), Bruce Means (Florida panhandle endemics), Jon Davenport (North Carolina specialities), Kent Bekker (*Ambystoma* unisexuals), Dan Gaillard (*Eurycea aquatica*), Greg George (reconnecting old contacts), and Rick Kneisel (providing Ohio information on *Ambystoma jeffersonianum*). Additionally, I thank both Jeff Beane and Alvin Braswell for providing information on *Eurycea arenicola* in North Carolina. I thank my former student, Mike Osbourn, for providing information and escorting me in the field in search of *Desmognathus organi*. I thank Chris Ogle and his colleagues Justin Talley, Jaci Baker, and Lillie Duffy for their assistance and escort in our search for *Plethodon pauleyi* in Tennessee. I thank Mark Mandica and the Amphibian Foundation for allowing me to photograph their *Ambystoma laterale*. Others who helped on some of our field excursions include Mark Watson (West Virginia), my Emory University herpetology class (Fall 2021), my son Kevin Wilson, and my two grandsons, Wyatt and Max Wilson. I finally thank Richard Montanucci for looking at and editing an early draft of our unisexual *Ambystoma* account.

I especially want to thank my wife, Dee, for her patience and understanding in the many days spent in the field looking for salamanders, nights out in the pouring rain, and endless days on the computer working on the manuscript and sorting images. ILYM.

PHOTO CREDITS

The authors thank the following individuals and organizations for providing photographs:

Brady Barr
Photograph on page 25 (right).

David Beamer
Photographs on pages 186 (middle), 339, 453 (top).

Jeff Beane
Photographs on pages 195, 196.

E. D. Brodie Jr.
Photographs on pages 24 (top left, bottom right), 25 (left), 83, 94 (bottom left), 277 (bottom), 392.

Robin Jung Brown
Photograph on page 459.

Kurt Buhlmann
Photograph on page 39 (top right).

Carlos Camp
Photographs on pages 5 (top), 163 (top), 181 (top).

E. D. Corey
Photograph on page 373 (top right).

Alan Cressler
Photographs on pages 3 (left), 36, 37, 38, 41 (right), 365.

David Dennis
Photographs on pages 3 (right), 11 (left), 13, 14 (bottom), 16 (bottom), 18 (top), 20 (bottom), 23, 24 (middle, bottom left), 27 (middle), 31, 55, 56 (bottom), 66, 67, 75, 80, 94 (top), 109, 117, 120 (middle), 138, 143 (top), 152 (bottom), 157 (middle), 162, 169 (bottom), 180, 181 (middle), 185, 187 (top), 189, 204, 205 (top), 214, 218, 219, 221 (bottom), 222, 240, 255, 261 (bottom), 262, 264, 275 (top), 285, 291 (bottom), 295 (right), 306 (top), 383 (top left, second from top, third from top), 406, 429, 436, 440 (right), 448 (bottom).

DigiMorph Co., University of Texas, Austin
Photograph on page 8.

Ken Dodd
Photographs on pages 63 (top), 206, 256 (bottom), 318 (left).

Sarah Eddy
Photograph on page 12.

Zach Felix
Photographs on pages 345, 403.

Dante Fenolio
Photographs on pages 119, 126 (top), 127, 128, 331, 334.

Don Forester
Photographs on pages 140, 216.

Mike Gibbons

Photograph on page 455 (top left).

Gabrielle Graeter

Photograph on page 9 (top left).

Mike Graziano

Photographs on pages i, 9 (bottom right), 49 (left, right), 53, 73, 74 (top, bottom), 104, 131, 200 (bottom), 213, 220, 259, 273, 277 (top), 279 (bottom), 287 (top), 305 (top), 306 (bottom), 307, 311 (top), 318 (right), 360 (top), 381, 382, 399, 422 (top), 444.

Cris Hagen

Photograph on page 30 (top).

Jeff Hall

Photographs on pages 76, 90 (bottom), 243 (bottom), 297 (top left), 449 (right).

Jesse Helton

Photograph on page 238.

Aubrey M. Heupel

Photograph on page 172.

E. Pierson Hill

Photographs on pages 59, 110 (top).

Kevin Hutcheson

Photographs on pages 87 (top), 114, 230 (second from top), 231 (top).

Jake Hutton

Photographs on pages 133, 239, 351, 353.

Vladimir Ischenko

Photograph on page 22.

John Jensen

Photographs on pages 4 (bottom), 68 (top left), 161, 250, 320, 373 (top left), 383 (top right), 419.

Trip Lamb

Photographs on pages 5 (bottom), 15, 159, 171, 287 (bottom), 315 (top), 387 (top).

Will Lattea

Photographs on pages 18 (bottom), 343, 346, 358, 378 (bottom right).

John MacGregor

Photographs on pages ii, 16 (top), 21 (bottom), 27 (top, bottom), 58 (left), 93, 136, 155, 168, 176 (top), 177 (bottom), 181 (bottom left), 187 (bottom), 258 (top), 263 (top), 270, 271 (top), 274 (right), 275 (bottom), 278, 292 (bottom), 308 (top), 310, 312 (fourth from top), 332, 361, 387 (bottom), 388 (middle), 393 (bottom), 401, 411 (top), 415 (bottom), 431, 432 (middle), 434, 448 (top),

Mike Marchand

Photograph on page 457 (bottom).

Johnathon Mays

Photographs on pages 163 (bottom), 167, 290, 295 (left top).

Kathryn Pawlik McCoard

Photographs on pages 177 (top), 202 (bottom).

Bruce Means

Photographs on pages 145 (top), 333.

Brian Miller

Photograph on page 432 (bottom).

Joe Mitchell

Photographs on pages v, 40 (left), 46, 209 (bottom), 215 (bottom), 388 (bottom left), 450 (left, right), 451 (bottom), 457 (top), 458 (top, bottom).

Brad Moon

Photographs on pages 14 (top), 19 (top),

20 (top), 21 (top), 43 (bottom), 56 (top), 105 (bottom right), 452, 453 (bottom).

Matt Niemiller
Photographs on pages 34, 105 (top), 181 (bottom right), 246 (top, middle).

Bill Peterman
Photographs on pages 61, 237, 388 (bottom right), 432 (top).

James Petranka
Photograph on page 315 (bottom).

Todd Pierson
Photograph on page 77.

Lisa Powers
Photographs on pages 4 (top), 17, 39 (top left), 134 (top), 217, 280, 284 (bottom), 460.

Todd Pusser
Photograph on page 56 (middle).

Steve Roble
Photograph on page 19 (bottom).

Francis Rose
Photograph on page 24 (top right).

Kevin Saunders
Photographs on pages 11 (right), 395 (left).

David Scott
Photographs on pages 314 (top), 383 (bottom left), 447 (left).

Jake Scott
Photographs on pages 85 (bottom), 107, 113, 244 (top), 335, 368.

Dirk Stevenson
Photographs on pages 44, 101, 142, 253, 274 (left), 281 (top), 294, 314 (bottom), 449 (left), 451 (top).

Steve Tilley
Photographs on pages 173 (bottom left, bottom right), 372, 418.

Brian Todd
Photographs on pages 7, 30 (bottom), 139 (top), 165, 209 (top), 447 (right).

Josef C. Uyeda
Photographs on pages 160 (top, bottom), 166 (bottom).

Robert Wayne Van Devender
Photographs on pages 9 (middle right), 40 (right), 58 (right), 60 (bottom), 64 (bottom), 70, 79, 81, 84, 85 (top, middle), 87 (bottom), 90 (top), 92, 98, 99 (top, middle, bottom), 105 (bottom left), 120 (top, middle), 125 (left, right), 134 (bottom), 143 (bottom), 145 (bottom), 146, 147, 153, 164, 169 (top), 173 (top left), 175 (top), 179, 186 (bottom), 197, 200 (top), 230 (second from bottom, bottom), 231 (bottom), 234, 256 (top), 271 (bottom), 281 (bottom), 284 (top), 289, 295 (left bottom), 312 (second from top, third from top, bottom right), 341, 344, 363, 367 (bottom), 369, 371, 373 (bottom left, bottom right), 378 (top, bottom left), 383 (fourth from top, third from bottom, second from bottom, bottom right), 386, 388 (top), 391, 395 (right), 396, 398, 408, 411 (bottom), 421, 422 (bottom), 424, 430, 441.

John White
Photographs on pages 241 (top), 271 (middle), 305 (bottom), 312 (top), 360 (bottom), 389.

Lori A. Williams
Photographs on pages 63 (bottom), 292 (top).

J. D. Willson

Photographs on pages 9 (bottom left), 68 (bottom), 110 (bottom), 112, 157 (top), 225 (bottom), 246 (bottom), 283 (top), 291 (top), 308 (bottom), 309, 312 (bottom left), 342, 375.

Larry Wilson

Photographs on pages vi, 9 (top right), 28 (top, bottom), 32 (left, right), 33 (left, right), 39 (bottom), 41 (left), 43 (top), 45 (top, bottom), 60 (top), 64 (top), 65, 68 (top right), 71, 89, 96, 102, 123, 124, 126 (bottom), 139 (bottom), 148, 149, 151, 152 (top), 156, 157 (bottom), 166 (top), 173 (top right), 175 (bottom), 176 (bottom), 182, 184, 186 (top), 191, 192, 193, 194, 199, 202 (top), 205 (middle, bottom), 207, 208, 210, 211, 215 (top), 221 (top), 223, 224, 225 (top), 226, 227, 228, 229, 230 (top), 235, 236, 241 (bottom), 243 (top), 244 (bottom), 245, 248, 249 (top, bottom left, bottom right), 251, 258 (bottom), 261 (top), 263 (bottom), 265, 266, 267 (top, bottom), 269, 279 (top), 286, 296, 297 (top right, bottom), 299 (top, bottom), 300 (top, bottom), 303, 304, 319, 323, 325 (top, bottom), 326, 327, 329, 330 (top, bottom), 337, 338, 340, 347, 349, 350, 352, 355, 359, 364, 367 (top), 376, 377, 385, 393 (top), 397, 400, 412, 414, 415 (top), 417, 420, 425, 428, 435, 438, 440 (left), 442, 455 (top right, bottom).

Robert Zappalorti

Photographs on pages 94 (bottom right), 283 (bottom), 298, 311 (bottom).

INDEX OF SCIENTIFIC NAMES

Page references in **bold** refer to species accounts. Page references in *italics* refer to illustrations not contained within a species account.

Plethodon shenandoah, 26, 350, **360–363**
Plethodon sherando, **377–381**
Plethodon shermani, **435–437**
Plethodon teyahalee, **418–420**
Plethodon variolatus, **382–388**
Plethodon ventralis, **431–434**
Plethodon virginia, **389–392**
Plethodon websteri, **425–427**
Plethodon wehrlei, 335, 338, 346, 351, 353, 354, **401–404**
Plethodon welleri, **405–408**
Plethodon yonahlossee, **364–367**
Pseudobranchus axanthus, **98–103**
Pseudobranchus axanthus axanthus, 100, 103
Pseudobranchus axanthus belli, 100, 103
Pseudobranchus striatus, **98–103**
Pseudobranchus striatus lustricolus, 103
Pseudobranchus striatus spheniscus, 103
Pseudobranchus striatus striatus, 103
Pseudotriton montanus, **240–244**
Pseudotriton montanus diastictus, 240, 244
Pseudotriton montanus flavissimus, 244
Pseudotriton montanus floridanus, 240, 244

Pseudotriton montanus montanus, 240, 244
Pseudotriton ruber, 15, **245–249**
Pseudotriton ruber nitidus, 249
Pseudotriton ruber ruber, 249
Pseudotriton ruber schencki, 249
Pseudotriton ruber vioscai, 249

Ranodon sibericus, 25

Salamandra salamandra, 25, 25
Salamandrella keyserlingii, 22, 22
Siren intermedia, **104–108**
Siren intermedia intermedia, 108
Siren intermedia nettingi. See Siren nettingi
Siren lacertina, **109–112**
Siren nettingi, **104–108**
Siren reticulata, **113–116**
Siren sphagnicola, **104–108**
Stereochilus marginatus, **73–76**

Typhlotriton spaeleus, 24

Urspelerpes brucei, 26, **250–252**

INDEX OF COMMON NAMES

Page references in **bold** refer to species accounts. Page references in *italics* refer to illustrations not contained within a species account.

dusky salamander (*continued*)
    identification of, 29, 30, 32, 153, 200,
        205, 219, 237, 356
    mountain, 28, 167, 175, 365
    northern, 21, 33, 144, 147, **172–79**,
        181, 187
    Pascagoula, **142–47**
    Piedmont, **172–79**
    as predator, 7
    as prey, 96, 238
    Santeetlah, **172–79**
    Savannah (Camp's), **172–79**
    southern, **142–47**, 169
    spotted, 144, 163, 169, **172–79**, 175
    Tilley's, **172–79**
    Valentine's southern, **142–47**
    western, **172–79**
    wolf, **172–79**
dwarf salamander, *3*, 228, **229–34**
    bog, **229–34**
    habitat of, 40, 42
    Hill's, **229–34**
    identification of, 30, 31, 73, 75,
        197, 227, 251, 306
    narrow-striped, *98*
    western, **229–34**
dwarf siren, 29, *30*
    broad-striped, 103
    Everglades, 100, 103
    Gulf hammock, 98, *99*, 103
    identification of, 105, 110, 114
    narrow-striped, *98*, *99*, 100, 103
    northern, **98–103**
    slender, 98, *99*, 103
    southern, **98–103**
European fire salamander, *23*, 25, *25*
European olm, 25, 26

fire salamander, 136
flat-headed salamander, **172–79**
flatwoods salamander, 44, 452
    frosted, *49*, **255–60**
    reticulated, **255–60**
four-toed salamander, 30, *40*, 41, 227,
    231, **305–10**

Georgia blind salamander, *37*, 45,
    119–22
giant salamander, 25, 103. *See also*
    hellbender
grampus. *See* hellbender
gray-cheeked salamander, 28
    Blue Ridge, **372–76**
    northern, *28*, **372–76**, 402, 422
    South Mountain, **372–76**
    southern, *28*, **372–76**
greater siren, *5*, 30, 56, 101, 105, **109–12**,
    114
green salamander, *4*, **323–26**, 329
    Blue Ridge Escarpment, 326, 330
    Hickory Nut Gorge, 323–24, 326,
        327–30, 370
    northern, 326, 330
    southern, 326, 330
grotto salamander, 23, *24*, 119

hellbender, **61–66**, *447*
    and biting, 10, *449*
    eastern, 66
    geographic distribution of, 8, 25
    habitat of, *40*, 41
    identification of, 26, 33, 94
    longevity of, 21
    Ozark, 66
    as predator, 13
    and reproduction, 18
imitator salamander, *131*, **148–51**, *448*

Japanese giant salamander, *25*, 26
Jefferson salamander, **261–65**
    identification of, 135, 266–67, 281,
        291, 295
    as predator, 309
    and unisexual offspring, 299–304
Jordan's salamander. *See* red-cheeked
    salamander
Junaluska salamander, 42, 192, **218–20**

lesser siren, 56, 101, **104–8**, 110, 114
    seepage, **104–8**
    western, **104–8**

long-tailed salamander, 45, 212, **213–17**
  identification of, 200, 205, 209, 222

Mabee's salamander, *9*, **286–89**
many-lined salamander, **73–76**
marbled salamander, *9, 34*, **290–93**, *447*,
  449, 452
  habitat of, 40
  identification of, 31, 134, 256–57, 262,
    271, 276, 281, 287, 294–95
  as predator, 14, 298, 309
  and reproduction, 17, 18
Mexican axolotl, *24*
Mexican mushroom salamander, *24*
mole salamander, *9*, 12, 15, 134–35, 256,
  270–74, 452
  identification of, 30, 256, 276–77,
    287, 291, 295
mole salamanders (Ambystomatidae),
  8, 17
  behavior of, 283
  habitat of, *38, 39*, 40, 42
  identification of, 32, 33, 81, 86, 261
  and reproduction, 255, 286, 292, 459
  size of, 280
  unisex, 266–67, 269, **299–304**
mollyhugger. *See* hellbender
mud devil. *See* hellbender
mud salamander, *15*, **240–44**
  eastern, 240, 244
  Gulf Coast, 244
  habitat of, *40, 41, 41*, 42
  identification of, 31, 75, 237, 246, 306
  midland, 240, 242, 244
  rusty, 240, 244
mudpuppy, *9*, 10, **93–97**, 97, 447, 454, 456
  geographic distribution of, 25
  habitat of, 37, 38, *40*
  identification of, 29, 30, 33, 62,
    85–86, 89
  Red River, 93, 95, *96*, 97
  and reproduction, 17–18

newt, 449
  activity of, 35

broken-striped, 312, 316, 319
central, 312, 316
defense mechanism of, 13–14
description of, 10
eastern, 14, 15, **311–16**
fire-bellied, 453
geographic distribution of, 25
habitat of, 37, *38*, 40–42
identification of, 30, 32
peninsula, 312, 316
as predator, 309
red-spotted, *9*, 14–15, 20–21, *40*, 298,
  312, 316
and reproduction, 20
striped, 31, 44, 312, **317–20**

Ocoee salamander, 31, 139, 148–50, **162–
  67**, 188, 356

patch-nosed salamander, 26, **250–52**
Peaks of Otter salamander, **339–42**, 347
Pigeon Mountain salamander, 45, **414–17**
pygmy salamander, **355–59**, 459
  habitat of, 44
  identification of, 26, 32, 139, 405, 407
  northern, **355–59**

ravine salamander, 390
  northern, **421–24**
  southern, 378, **421–24**
red eft. *See* newt
Red Hills salamander, **331–34**, 350
red salamander, **245–49**
  black-chinned, 245, *248*, 249
  Blue Ridge, 245, *248*, 249
  defense mechanism of, 14–15
  habitat of, *41*, 42
  identification of, 31, 237, 306
  northern, *245*, 249
  as predator, 171, 178, 207
  as prey, 238
  southern, 245, *248*, 249
red-backed salamander, 362, 372, 376, *448*
  eastern, *27, 321*, 348, 373, **377–81**,
    385, 390, 399, 400, 405

Big Mouth Cave salamander, 126
    pinkish cave salamander, 126
three-lined salamander, *32*, **208–12**,
    215–17
    habitat of, *41*, 42
    identification of, 31, 200, 205
tiger salamander, *9*, **280–85**, 447, 452,
    453, *455*
    activity of, 10
    barred, 285
    defense mechanism of, 13
    habitat of, *39*, 44, 46
    as predator, 7, 12, 278, 288
    as prey, 14
    and reproduction, 34
    and unisexual offspring, 299–304
two-lined salamander, 220, 250
    Blue Ridge, 21, 191, 192, 200, **204–7**,
    219
    habitat of, *41*, 42, 46
    identification of, 31, 188, 251, 306
    northern, **199–203**, 207
    as prey, 70, 96, 238, 243
    southern, 32, 191–94, 197, 198, **199–
    203**, 205, 207, 227, 231

Valley and Ridge salamander, 28, 348,
    378, **389–92**

waterdog, 10, 30, 454, 456
    Apalachicola River, **84–88**, 89
    Black Warrior (Alabama), **77–80**, 89, 94
    dwarf, 81, 86, **89–92**
    Escambia River, **84–88**, 89
    geographic distribution of, 25
    Gulf Coast, 34, 80, **84–88**, 89
    habitat of, 37, 41
    Neuse River, **81–83**, 89, 90
Webster's salamander, 34–35, 399, **425–
    27**, 432, 434
Wehrle's salamander, 335, 338, 344, 346,
    401–4
    identification of, 336, 345, 353–54,
    385, 422, 442
Weller's salamander, **405–8**, 422
woodland salamander. *See Plethodon species
    in index of scientific names*
woodland salamander, yellow-spotted,
    351–54

Yonahlossee salamander, 5, *18*, **364–67**,
    369, 371, 373, 407

zigzag salamander, 31, 45, 425–26, 427
    northern, 422, **431–34**
    Ozark, 434
    southern, 399, **431–34**